Plant Genes, Genomes and Genetics

Plant Genes, Genomes and Genetics

Kevin Sterling

STATES
ACADEMIC PRESS
www.statesacademicpress.com

States Academic Press,
109 South 5th Street,
Brooklyn, NY 11249, USA

Visit us on the World Wide Web at:
www.statesacademicpress.com

ISBN: 978-1-63989-421-5 (Hardback)

Cataloging-in-Publication Data

Plant genes, genomes and genetics / Kevin Sterling.
p. cm.
Includes bibliographical references and index.
ISBN 978-1-63989-421-5
1. Plant genetics. 2. Plant gene expression. 3. Plant genetic regulation.
4. Plant genomes. I. Sterling, Kevin.

QK981 .P53 2022
581.35--dc23

TABLE OF CONTENTS

PREFACE

It is with great pleasure that I present this book. It has been carefully written after numerous discussions with my peers and other practitioners of the field. I would like to take this opportunity to thank my family and friends who have been extremely supporting at every step in my life.

The study of genes, heredity and genetic variation in plants is known as plant genetics. It is the sub-field of biology and botany. The field aims to understand the patterns through which traits are passed down by parents to offsprings. The Deoxyribonucleic acids (DNA) is the nucleic acid which carries long-term storage of information and contains the information needed to make other components of cell. The Deoxyribonucleic acids (DNA) in plants carry the genetic information of the plants which are called genes. The field focuses on understanding the role that different genes perform within a genome. This information is further used for plant breeding and manipulation of plant genes to produce various other desired genotypes. This book provides comprehensive insights into the field of plant genetics. It is compiled in such a manner, that it will provide an in-depth knowledge about the theory and practice of this discipline. As this field is emerging at a rapid pace, the contents of this book will help the readers understand the modern applications of the subject.

The chapters below are organized to facilitate a comprehensive understanding of the subject:

Chapter - Plant Genes

Plant genes refer to the units of inheritance that are acquired from parent plant by its offspring which form the basis of plant genetics. Types of genes include abiotic stress-inducible genes, transcriptional factor genes, etc. This chapter sheds light on the subject of plant genes and its related aspects for a thorough understanding of the subject.

Chapter - Plant Genomics

Plant genomics deals with the study of genetic compositions, structures, sequencing functions and interactions of an entire plant genome. A few of its aspects are plant genome assembly, microRNA sequencing, genome reference sequencing and resequencing, etc. This chapter will briefly introduce about plant genomics.

Chapter - Plant Genetics

Plants genetics is an interdisciplinary field that studies genes, genetic variation and heredity in plants. It further includes the study of plant genetic resources, crop plant evolution and doubled haploids. This chapter has been carefully written to provide an easy understanding of plant genetics.

Chapter - Plant Reproductive Systems

Plant reproductive system is defined as the system by which plants reproduce. It is branched into two types - sexual and asexual reproduction. In sexual reproduction the offspring is produced by the fusion of gametes. In asexual reproduction the offspring is produced without the fusion of gametes. The topics elaborated in this chapter will help in gaining a better perspective of the subject of plant reproductive system.

Chapter - Plant Breeding

The science of altering traits of plants to create desired genotypes and phenotypes for improving the genetic potential of plants is termed as plant breeding. It involves controlled pollination and genetic engineering followed by artificial selection of progeny. This chapter delves into plant breeding for an in-depth understanding of the subject.

Kevin Sterling

Plant Genes

1 CHAPTER

Plant genes refer to the units of inheritance that are acquired from parent plant by its offspring which form the basis of plant genetics. Types of genes include abiotic stress-inducible genes, transcriptional factor genes, etc. This chapter sheds light on the subject of plant genes and its related aspects for a thorough understanding of the subject.

LEAFY

The LEAFY (LFY) protein is a key regulator of flower development in angiosperms. Its gradually increased expression governs the sharp floral transition, and LFY subsequently controls the patterning of flower meristems by inducing the expression of floral homeotic genes. Despite a wealth of genetic data, how LFY functions at the molecular level is poorly understood. Crystal structures for the DNA-binding domain of Arabidopsis thaliana LFY bound to two target promoter elements. LFY adopts a novel seven-helix fold that binds DNA as a cooperative dimer, forming base-specific contacts in both the major and minor grooves. Cooperativity is mediated by two basic residues and plausibly accounts for LFY's effectiveness in triggering sharp developmental transitions.

Homeotic genes control developmental patterns and organ morphogenesis. In animals, they encode transcription factors of the homeodomain family, such as Hox and paired proteins, which contact DNA through one or several helix-turn-helix (HTH) motifs In plants, most homeotic genes determining the identity of floral organs encode MADS-box transcription factors, suggesting that plants and animals have adopted distinct types of homeotic regulators. In addition to organ identity genes, plants also use another class of regulators named 'meristem identity genes', which control floral meristem versus shoot/inflorescence fate. In Arabidopsis thaliana, the meristem identity genes LEAFY (LFY) and APETALA1 (AP1) induce flower development, whereas TERMINAL FLOWER1 (TFL1) promotes inflorescence development. Mutations or ectopic expression of these genes result in complete or partial interconversions between flower and inflorescence meristems.

The LFY gene encodes a plant-specific transcription factor, which has a cardinal function in this process, regulating both the transition to flowering and the subsequent patterning of young floral meristems. During the plant vegetative growth, LFY expression increases in newly formed leaves until a certain threshold is reached. LFY then induces the expression of AP1 and CAULIFLOWER (CAL) genes and triggers the abrupt floral transition.Once the floral meristem is established, LFY governs its spatial patterning by inducing the expression of the floral homeotic ABC genes, such as AP1, AP3 or AGAMOUS (AG), which control the identity of stereotypically arranged floral organs.

LFY is found in all terrestrial plants from moss to angiosperms; its sequence shows a high level of conservation throughout the plant kingdom but no apparent similarity to other proteins Unlike many plant transcription factors that evolved by gene duplication to form a multigene family, LFY is present in single copy in most angiosperms and lfy mutants available from several species such as snapdragon, petunia, tomato or maize show, as in Arabidopsis, partial or complete flower-to-shoot conversions. In gymnosperms, a paralogous NEEDLY (NLY) clade of genes exists. No mutant is available in these species, but LFY and NLY expression patterns are also consistent with a role in reproductive organ development. Because of its central role in determining floral meristem identity, and considering that NLY disappeared concomitantly with the appearance of flowers, LFY has been put at the centre of different evolutionary scenarios that rationalize the appearance of the successful angiosperm group.

LFY activates gene expression by recognizing pseudo-palindromic sequence elements (CCANTGT/G) in the promoters of its target genes, including AP1 (one site) and AG (four sites; AG-I to AG-IV) LFY has two domains, a partially conserved N-terminal domain that is thought to contribute to transcriptional activation and a highly conserved C-terminal domain responsible for DNA binding (LFY-C). LFY functions synergistically with coregulators such as the WUSCHEL (WUS) homeodomain protein or the UFO F-Box protein.

AC/DS TRANSPOSABLE CONTROLLING ELEMENTS

Barbara McClintock reported that the Ac/Ds transposable element system can generate major chromosomal rearrangements (MCRs), but the underlying mechanism has not been determined. Here, we identified a series of chromosome rearrangements derived from maize lines containing pairs of closely linked Ac transposable element termini. Molecular and cytogenetic analyses showed that the MCRs in these lines comprised 17 reciprocal translocations and two large inversions. The breakpoints of all 19 MCRs are delineated by Ac termini and characteristic 8-base-pair target site duplications, indicating that the MCRs were generated by precise transposition reactions involving the Ac termini of two closely linked elements. This alternative transposition mechanism may have contributed to chromosome evolution and may also occur during V(D)J recombination resulting in oncogenic translocations.

In the 1940s, Barbara McClintock reported that the maize Activator (Ac) element could induce transposition of the nonautonomous Dissociation (Ds) element, which she identified as a locus of chromosome breakage. She also showed, using cytogenetic methods, that transposition of Ds was sometimes accompanied by major chromosomal rearrangements (MCRs) including deletions, duplications, inversions, reciprocal translocations, and ring chromosomes. Because these rearrangements only occurred in the presence of Ac and they retained Ds at their breakpoints, McClintock concluded that Ac/Ds transposition is somehow responsible for their origin. Although several hypotheses have been advanced to explain the origin of McClintock's MCRs, it is still unclear precisely how these large chromosome rearrangements were generated.

Standard Ac/Ds transposition only changes the position of the transposon in the genome, hence the MCRs isolated by McClintock must have originated by some other mechanism involving transposition. A type of aberrant Ds transposition that results in fusion of sister chromatids, chromosome breakage, and formation of deletions has been described. Previously, we showed that a pair of Ac termini in tandem orientation produced reciprocal deletion/duplication alleles via a mechanism in which Ac transposase interacts with Ac termini on sister chromatids; thus it was termed sister chromatid transposition (SCT). However, SCT cannot fully explain the origin of McClintock's MCRs because the only heritable products it generates are deletions and inverted duplications.

Recently, we and others showed that a pair of Ac termini in reversed orientation can also undergo transposition (reversed Ac ends transposition). In this reaction, Ac transposase acts on a pair of Ac termini on the same sister chromatid.Insertion of the excised Ac termini into nearby sites can produce relatively small rearrangements, including deletions and inversions. However, the excised Ac ends could also insert into distant sites on the same or different chromosomes to generate a variety of MCRs. For example, insertion into the same chromatid would generate an acentric fragment and a ring chromosome (the top portion of; Supplemental Movie 1), or an inversion (the bottom portion of ; Supplemental Movie 1), depending on the orientation of insertion of the transposon ends. In addition, insertion into another chromosome would generate a reciprocal translocation; Supplemental Movie 1), or a dicentric chromosome and an acentric fragment (Supplemental Movie 1), again depending on the orientation of insertion of the transposon ends.

The Ac Activator element is autonomous, whereas the Ds Dissociation element requires an Activator element to transpose. Ac was initially discovered as enabling a Ds element to break chromosomes. Both Ac and Ds can also insert into genes, causing mutants that may revert to normal on excision of the element. The phenotypic consequence of Ac/Ds transposable element includes mosaic colors in kernels and leaves in maize.

Its discovery was based on studying its genetic behavior, i.e., "jumping genes" in maize and published by Barbara McClintock, leading to her 1983 Nobel Prize in Medicine.

The Ac/Ds transposable elements were first isolated and sequenced By Federoff et al. 1983 using insertions of Ac and Ds into the well-studied Waxy(Wx1) gene. The elements have been shown to function in other plants, including tobacco, Arabidopsis,) and rice.

Genomic analysis of maize show that these elements, which share terminal 11 bp imperfect inverted repeat sequences, have much sequence heterogeneity, both in length and content. They also include a class of DNA elements that do not transpose in the presence of the Ac element. The chromosome breaking property has been shown to come from pairs of closely positioned elements.

Toolkits

Researchers use mutant phenotypes to discover gene functions. Ac/Ds prefer to transpose to nearby genes, affording a way to mutagenize those regions of the genome, and by subsequent genetic crosses, remove the Ac that causes new mutants and instability of a Ds mutant. Collections of Ac/Ds elements that cover the genome and are useful for generating mutants. Application of Ac/Ds toolkits has also been applied to other species like arabidopsis, yeast, and even zebrafish.

In Corn

Activator (Ac)/Dissociation (Ds) transposable elements were discovered by Barbara McClintock when she was studying the maize genomic composition of the short arm of chromosome 9. She noticed that when chromosome 9 had been exposed to drastic cultural modifications, the progeny had changes such as multiple copies of the short arm or lacking one or more of its parts, as well as other changes. She believed that these changes were due to transposition of "mutable loci" into the genome and that these spontaneous translocations were not random due to where the breaks occurred and where they fused.

Ac/Ds elements have been observed to insert into gene rich regions of the maize genome, they alter the regulation of gene expression and may create unstable insertion alleles, stable derivatives, or excision alleles due to insertion of a transposable element into a gene. Transposable elements residing at or near a gene prevent gene expression and can also result in a mutation that causes exhibition of the recessive phenotype. Removal of transposable element locus results in restoration of the gene organization and activity. Ac is 4565 base pairs long and codes for a 3.5 kb open reading frame that synthesizes an 807 amino acid long transposase enzyme. Ac elements are autonomous and their movement results in a 4.3 kb insertion. Ds elements are not autonomous because they cannot produce the transposase needed for transposition, and can only transpose when it is provided by the Ac element. Ds elements have shown to cause a 4.1kB and 2.0 kB insertions. The transposable elements were seen in progeny of plants that had undergone stress, and mutations caused by the insertion are like those caused by x-rays, UV light, or chemicals causing events like chromosome breakage and fusion.

There are distinct families of transposon controlling elements that are made up of a combination of elements some that can, and some that cannot, transpose. The families differ in their developmental timing and transposition frequency, as well other types of genetic rearrangements. Ac or Ds element insertion near a locus causes unstable mutations. Ds elements are considered non-autonomous because they cannot transpose without the presence of the Ac element, since they themselves cannot produce the enzyme transposase needed for transposition. However, since Ds elements can utilize the transposase enzyme produced by the Ac elements, it shows that they have the structural information needed for transposition and only lack the information needed to produce the enzyme. Ds elements were identified at the site of a chromosome breakage. Ac and Ds elements are structurally related because insertion of either element brings about similar mutations. They are also similar because their restriction endonuclease cleavage site maps are indistinguishable from each other.

Different mutants display different levels of gene expression, which largely depends on the presence or absence of the transposable element somewhere else in the genome. The different functions of the elements are detected as altered temporal or spatial patterns of somatic reversions or reversible inactivation of the entire element. Some excision events of these elements restore gene function and can be detected as somatic reversions. If Ac or Ds insertions in an exon, and the transposable elements are excised leaving some of the duplicated base pairs behind, it alters the protein structure either by causing mutations such as frame shift mutations or the addition of amino acids. In some cases, unstable Ds or Ac induced mutant can give rise to a stable recessive mutant. Ac determines the mutation process and the mutable loci as well as their timing. Ac transposase does not influence transcription initiation site selection, and large numbers of Ac elements may inhibit the expression of Ds by reducing the rate of transcription initiation instead of affecting the transcription site selection. Ac transposase is also capable of suppressing gene expression when Ac or Ds inserts in 5' untranslated region using target sites. Ac or Ds element insertion sites have been characterized by the presence of different direct duplications of 6-10 base pairs prior to insertions, indicating that Ac transposase may have preference for short duplication as insertion sites. Transposition of both Ac and Ds occurs during development of a tissue and is under precise control which is determined by the number of Ac loci present, their organization, and their position in the chromosome complement. The transposable elements not only entirely remove or alter the gene function via insertion, but can also exert a mutator activity when they leave the position where they had visited the chromosome. Induced disturbances in quantity and organization of the heterochromatic elements of the chromosome could give rise to a series of alterations in its structure, behavior, and in genic reactions that can alter phenotypic expression.

The Ac and Ds elements share two properties that are common for transposable elements. One is that at the site of their insertion, a short DNA sequence is 8 base pairs long, and is duplicated and borders exactly at the Ac or Ds sequence. Second, the sequences of several Ds elements terminate in an inverted repeat of the nucleotide sequence TAGGATGAAA. The Ds elements shows quite some degree of similarity with the Ac elements, such as the Ds9 element which is a complemental mutant of Ac and only differs in the open reading frame 1. Other Ds elements differ from the Ac element by internal deletions.

As Ac copy number increases, the expression of genes flanking the Ds promoter insertion are negatively regulated, which is called the negative dosage effect. Ds flanking gene expression is inhibited in the presence of the autonomous Ac element. As the Ac element copy number increases, transposition events that require Ac occur at a later time during endosperm development. Ac causes no or few chromosome breaks. In the absence of Ac, Ds caused mutations are stable. By identifying alleles that suppress the Ac element they could be used for functional genomics studies. The properties of transposable elements can also be altered in a clonally heritable fashion. Ds elements can be used as building blocks for complicated structures like the double Ds or the 30kb transposon-like insertion which is terminated by Ds elements. Ds elements can generate long direct inverted duplicates of a chromosome segment, indicating that Ds elements are able to mobilize DNA sequences unrelated to themselves, this can provide a mechanism for rearrangement of genetic information. Since small duplication of host DNA created by the insertion of the elements are left over in a slightly altered form after excision of the transposable element, they may indicate that plant transposable elements play a role in the evolution of genes.

CHLOROPLAST DNA

The chloroplast genome consists of homogeneous circular DNA molecules. To date, the entire nucleotide sequences (120-190 kbp) of chloroplast genomes have been determined from eight plant species. The chloroplast genomes of land plants and green algae contain about 110 different genes, which can be classified into two main groups: genes involved in gene expression and those related to photosynthesis. The red alga Porphyra chloroplast genome has 70 additional genes, one-third of which are related to biosynthesis of amino acids and other low molecular mass compounds. Chloroplast genes contain at least three structurally distinct promoters and transcribe two or more classes of RNA polymerase. Two chloroplast genes, rps12 of land plants and psaA of Chlamydomonas, are divided into two to three pieces and scattered over the genome. Each portion is transcribed separately, and two to three separate transcripts are joined together to yield a functional mRNA by trans-splicing. RNA editing (C to U base changes) occurs in some of the chloroplast transcripts. Most edited codons are functionally significant, creating start and stop codons and changing codons to retain conserved amino acids.

Chloroplasts have their own DNA, often abbreviated as cpDNA. It is also known as the plastome when referring to genomes of other plastids. Its existence was first proven in 1962, and first sequenced in 1986 when two Japanese research teams sequenced the chloroplast DNA of liverwort and tobacco. Since then, hundreds of chloroplast DNAs from various species have been sequenced, but they are mostly those of land plants and green algae glaucophytes, red algae, and other algae groups are extremely under-represented, potentially introducing some bias in views of "typical" chloroplast DNA structure and content.

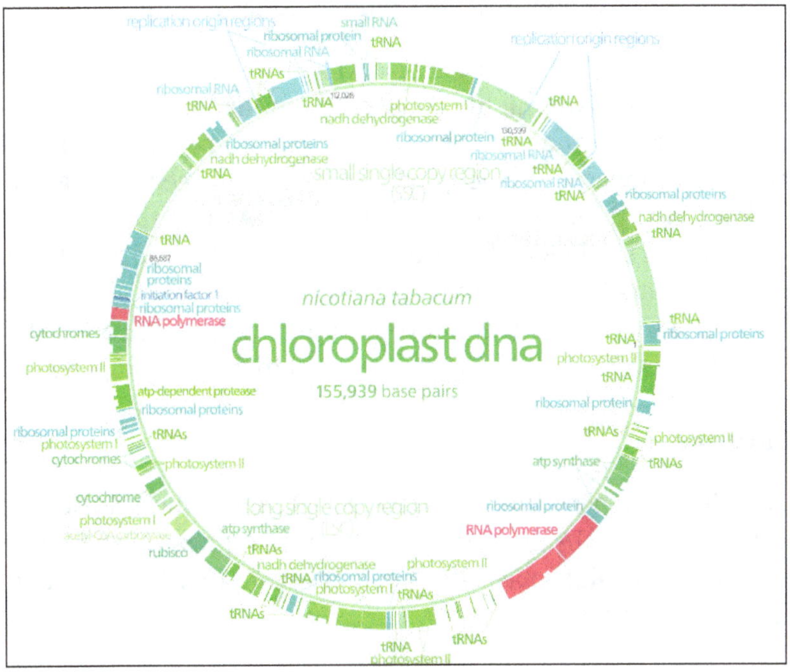

Chloroplast DNA Interactive gene map of chloroplast DNA from Nicotiana tabacum. Segments with labels on the inside reside on the B strand of DNA, segments with labels on the outside are on the A strand. Notches indicate introns.

Molecular Structure

Chloroplast DNAs are circular, and are typically 120,000–170,000 base pairs long. They can have a contour length of around 30–60 micrometers, and have a mass of about 80–130 million daltons.

Most chloroplasts have their entire chloroplast genome combined into a single large ring, though those of dinophyte algae are a notable exception their genome is broken up into about forty small plasmids, each 2,000–10,000 base pairs long. Each minicircle contains one to three genes, but blank plasmids, with no coding DNA, have also been found.

Inverted Repeats

Many chloroplast DNAs contain two inverted repeats, which separate a long single copy section (LSC) from a short single copy section (SSC).

The inverted repeats vary wildly in length, ranging from 4,000 to 25,000 base pairs long each. Inverted repeats in plants tend to be at the upper end of this range, each being 20,000–25,000 base pairs long. The inverted repeat regions usually contain three ribosomal RNA and two tRNA genes, but they can be expanded or reduced to contain as few as four or as many as over 150 genes. While given pair of inverted repeats are rarely completely identical, they are always very similar to each other, apparently resulting from concerted evolution.

The inverted repeat regions are highly conserved among land plants, and accumulate few mutations. Similar inverted repeats exist in the genomes of cyanobacteria and the other two chloroplast lineages (glaucophyta and rhodophyceæ), suggesting that they predate the chloroplast, though some chloroplast DNAs like those of peas and a few red algae have since lost the inverted repeats. Others, like the red alga Porphyra flipped one of its inverted repeats (making them direct repeats). It is possible that the inverted repeats help stabilize the rest of the chloroplast genome, as chloroplast DNAs which have lost some of the inverted repeat segments tend to get rearranged more.

Linear Structure

Chloroplast DNA has long been thought to have a circular structure, but some evidence suggests that chloroplast DNA more commonly takes a linear shape. Over 95% of the chloroplast DNA in corn chloroplasts has been observed to be in branched linear form rather than individual circles.

Nucleoids

Each chloroplast contains around 100 copies of its DNA in young leaves, declining to 15–20 copies in older leaves. They are usually packed into nucleoids which can contain several identical chloroplast DNA rings. Many nucleoids can be found in each chloroplast.

Though chloroplast DNA is not associated with true histones, in red algae, a histone-like chloroplast protein (HC) coded by the chloroplast DNA that tightly packs each chloroplast DNA ring into a nucleoid has been found.

In primitive red algae, the chloroplast DNA nucleoids are clustered in the center of a chloroplast, while in green plants and green algae, the nucleoids are dispersed throughout the stroma.

DNA Replication

Leading Model of cpDNA Replication

The mechanism for chloroplast DNA (cpDNA) replication has not been conclusively determined, but two main models have been proposed. Scientists have attempted to observe chloroplast replication via electron microscopy since the 1970s. The results of the microscopy experiments led to the idea that chloroplast DNA replicates using a double displacement loop (D-loop). As the D-loop moves through the circular DNA, it adopts a theta intermediary form, also known as a Cairns replication intermediate, and completes replication with a rolling circle mechanism. Replication starts at specific points of origin. Multiple replication forks open up, allowing replication machinery to replicate the DNA. As replication continues, the forks grow and eventually converge. The new cpDNA structures separate, creating daughter cpDNA chromosomes.

In addition to the early microscopy experiments, this model is also supported by the amounts of deamination seen in cpDNA. Deamination occurs when an amino group is lost and is a mutation that often results in base changes. When adenine is deaminated, it becomes hypoxanthine. Hypoxanthine can bind to cytosine, and when the XC base pair is replicated, it becomes a GC (thus, an A → G base change).

Original DNA Strand

...CCATGCATGGATC...

Deamination of an Adenine

...CCATGCATGGATC...

↓

...CCHTGCATGGATC...

During Replication, H pairs with C

...CCHTGCATGGATC...
...GGCACGTACCTAG...

When Replicated Again, C pairs with G

...GGCACGTACCTAG...
...CCGTGCATGGATC...

Over time, base changes in the DNA sequence can arise from deamination mutations. When adenine is deaminated, it becomes hypoxanthine, which can pair with cytosine. During replication, the cytosine will pair with guanine, causing an A → G base change.

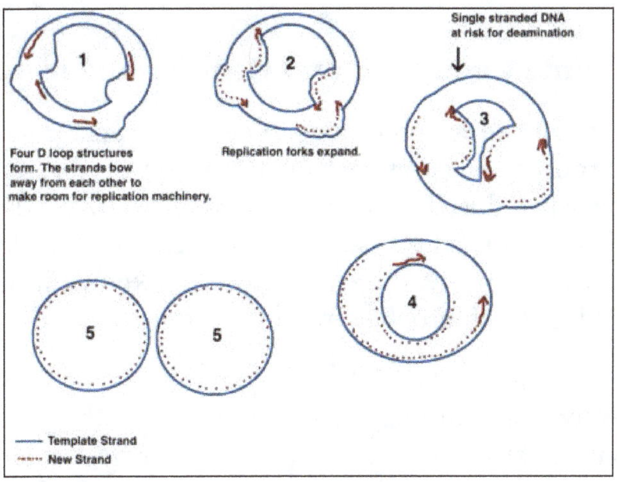

Chloroplast DNA replication via multiple D loop mechanisms.

Alternative Model of Replication

One of the main competing models for cpDNA asserts that most cpDNA is linear and participates in homologous recombination and replication structures similar to bacteriophage T4. It has been established that some plants have linear cpDNA, such as maize, and that more still contain complex structures that scientists do not yet understand; however, the predominant view today is that most cpDNA is circular. When the original experiments on cpDNA were performed, scientists did notice linear structures; however, they attributed these linear forms to broken circles. If the branched and complex structures seen in cpDNA experiments are real and not artifacts of concatenated circular DNA or broken circles, then a D-loop mechanism of replication is insufficient to explain how those structures would replicate. At the same time, homologous recombination does not explain the multiple A → G gradients seen in plastomes. This shortcoming is one of the biggest for the linear structure theory.

Gene Content and Protein Synthesis

The chloroplast genome most commonly includes around 100 genes which code for a variety of things, mostly to do with the protein pipeline and photosynthesis. As in prokaryotes, genes in chloroplast DNA are organized into operons. Introns are common in chloroplast DNA molecules, while they are rare in prokaryotic DNA molecules (plant mitochondrial DNAs commonly have introns, but not human mtDNA).

Among land plants, the contents of the chloroplast genome are fairly similar they code for four ribosomal RNAs, 30–31 tRNAs, 21 ribosomal proteins, and four RNA polymerase subunits, involved in protein synthesis. For photosynthesis, the chloroplast DNA includes genes for 28 thylakoid proteins and the large Rubisco subunit. In addition, its genes encode eleven subunits of a protein complex which mediates redox reactions to recycle electrons, which is similar to the NADH dehydrogenase found in mitochondria.

Chloroplast Genome Reduction and Gene Transfer

Over time, many parts of the chloroplast genome were transferred to the nuclear genome of the host, a process called endosymbiotic gene transfer. As a result, the chloroplast genome is heavily reduced compared to that of free-living cyanobacteria. Chloroplasts may contain 60–100 genes whereas cyanobacteria often have more than 1500 genes in their genome. Contrarily, there are only a few known instances where genes have been transferred to the chloroplast from various donors, including bacteria.

Endosymbiotic gene transfer is how we know about the lost chloroplasts in many chromalveolate lineages. Even if a chloroplast is eventually lost, the genes it donated to the former host's nucleus persist, providing evidence for the lost chloroplast's existence. For example, while diatoms (a heterokontophyte) now have a red algal derived chloroplast, the presence of many green algal genes in the diatom nucleus provide evidence that the diatom ancestor (probably the ancestor of all chromalveolates too) had a green algal derived chloroplast at some point, which was subsequently replaced by the red chloroplast.

In land plants, some 11–14% of the DNA in their nuclei can be traced back to the chloroplast, up to 18% in Arabidopsis, corresponding to about 4,500 protein-coding genes. There have been a few recent transfers of genes from the chloroplast DNA to the nuclear genome in land plants.

Proteins Encoded by the Chloroplast

Of the approximately three-thousand proteins found in chloroplasts, some 95% of them are encoded by nuclear genes. Many of the chloroplast's protein complexes consist of subunits from both the chloroplast genome and the host's nuclear genome. As a result, protein synthesis must be coordinated between the chloroplast and the nucleus. The chloroplast is mostly under nuclear control, though chloroplasts can also give out signals regulating gene expression in the nucleus, called retrograde signaling.

Protein Synthesis

Protein synthesis within chloroplasts relies on an RNA polymerase coded by the chloroplast's own genome, which is related to RNA polymerases found in bacteria. Chloroplasts also contain a mysterious second RNA polymerase that is encoded by the plant's nuclear genome. The two RNA polymerases may recognize and bind to different kinds of promoters within the chloroplast genome. The ribosomes in chloroplasts are similar to bacterial ribosomes.

RNA Editing in Plastids

RNA editing is the insertion, deletion, and substitution of nucleotides in a mRNA transcript prior to translation to protein. The highly oxidative environment inside chloroplasts increases the rate of mutation so post-transcription repairs are needed to conserve functional sequences. The chloroplast editosome substitutes $C \rightarrow U$ and $U \rightarrow C$ at very specific locations on the transcript. This can change the codon for an amino acid or restore a non-functional pseudogene by adding an AUG start codon or removing a premature UAA stop codon.

The editosome recognizes and binds to cis sequence upstream of the editing site. The distance between the binding site and editing site varies by gene and proteins involved in the editosome. Hundreds of different PPR proteins from the nuclear genome are involved in the RNA editing process. These proteins consist of 35-mer repeated amino acids, the sequence of which determines the cis binding site for the edited transcript.

Basal land plants such as liverworts, mosses and ferns have hundreds of different editing sites while flowering plants typically have between thirty and forty. Parasitic plants such as Epifagus virginiana show a loss of RNA editing resulting in a loss of function for photosynthesis genes.

Protein Targeting and Import

The movement of so many chloroplast genes to the nucleus means that many chloroplast proteins that were supposed to be translated in the chloroplast are now synthesized in the cytoplasm. This means that these proteins must be directed back to the chloroplast, and imported through at least two chloroplast membranes.

Curiously, around half of the protein products of transferred genes aren't even targeted back to the chloroplast. Many became exaptations, taking on new functions like participating in cell division, protein routing, and even disease resistance. A few chloroplast genes found new homes in the mitochondrial genome most became nonfunctional pseudogenes, though a few tRNA genes still work in the mitochondrion. Some transferred chloroplast DNA protein products get directed to the secretory pathway (though many secondary plastids are bounded by an outermost membrane derived from the host's cell membrane, and therefore topologically outside of the cell, because to reach the chloroplast from the cytosol, you have to cross the cell membrane, just like if you were headed for the extracellular space. In those cases, chloroplast-targeted proteins do initially travel along the secretory pathway).

Because the cell acquiring a chloroplast already had mitochondria (and peroxisomes, and a cell membrane for secretion), the new chloroplast host had to develop a unique protein targeting system to avoid having chloroplast proteins being sent to the wrong organelle.

Cytoplasmic Translation and N-terminal Transit Sequences

Polypeptides, the precursors of proteins, are chains of amino acids. The two ends of a polypeptide are called the N-terminus, or amino end, and the C-terminus, or carboxyl end. For many (but not all) chloroplast proteins encoded by nuclear genes, cleavable transit peptides are added to the N-termini of the polypeptides, which are used to help direct the polypeptide to the chloroplast for import (N-terminal transit peptides are also used to direct polypeptides to plant mitochondria). N-terminal transit sequences are also called presequences because they are located at the "front" end of a polypeptide ribosomes synthesize polypeptides from the N-terminus to the C-terminus.

Chloroplast transit peptides exhibit huge variation in length and amino acid sequence. They can be from 20–150 amino acids long an unusually long length, suggesting that transit peptides are actually collections of domains with different functions. Transit peptides tend to be positively charged, rich in hydroxylated amino acids such as serine, threonine, and proline, and poor in acidic amino acids like aspartic acid and glutamic acid. In an aqueous solution, the transit sequence forms a random coil.

Not all chloroplast proteins include a N-terminal cleavable transit peptide though. Some include the transit sequence within the functional part of the protein itself. A few have their transit sequence appended to their C-terminus instead. Most of the polypeptides that lack N-terminal targeting sequences are the ones that are sent to the outer chloroplast membrane, plus at least one sent to the inner chloroplast membrane.

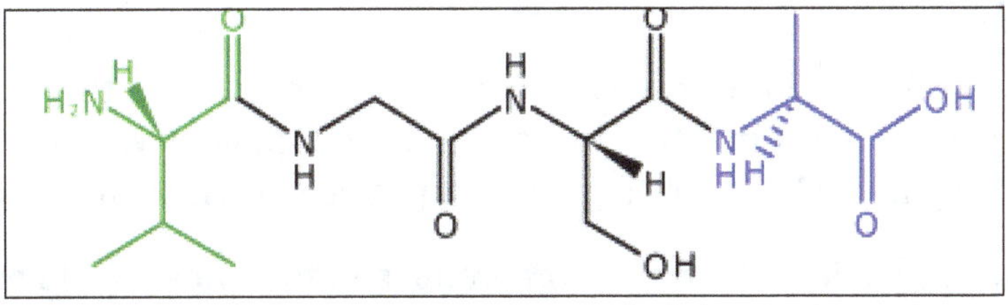

A polypeptide with four amino acids linked together. At the left is the N-terminus, with its amino (H_2N) group in green. The blue C-terminus, with its carboxyl group (CO_2H) is at the right.

Phosphorylation, Chaperones and Transport

After a chloroplast polypeptide is synthesized on a ribosome in the cytosol, ATP energy can be used to phosphorylate, or add a phosphate group to many (but not all) of them in their transit sequences. Serine and threonine (both very common in chloroplast transit sequences making up 20–30% of the sequence) are often the amino acids that accept the phosphate group. The enzyme that carries out the phosphorylation is specific for chloroplast polypeptides, and ignores ones meant for mitochondria or peroxisomes.

Phosphorylation changes the polypeptide's shape, making it easier for 14-3-3 proteins to attach to the polypeptide. In plants, 14-3-3 proteins only bind to chloroplast preproteins. It is also bound by the heat shock protein Hsp70 that keeps the polypeptide from folding prematurely. This is important because it prevents chloroplast proteins from assuming their active form and carrying out their chloroplast functions in the wrong place the cytosol. At the same time, they have to keep just enough shape so that they can be recognized and imported into the chloroplast.

The heat shock protein and the 14-3-3 proteins together form a cytosolic guidance complex that makes it easier for the chloroplast polypeptide to get imported into the chloroplast.

Alternatively, if a chloroplast preprotein's transit peptide is not phosphorylated, a chloroplast preprotein can still attach to a heat shock protein or Toc159. These complexes can bind to the TOC complex on the outer chloroplast membrane using GTP energy.

The Translocon on the Outer Chloroplast Membrane (TOC)

The TOC complex, or translocon on the outer chloroplast membrane, is a collection of proteins that imports preproteins across the outer chloroplast envelope. Five subunits of the TOC complex have been identified two GTP-binding proteins Toc34 and Toc159, the protein import tunnel Toc75, plus the proteins Toc64 and Toc12.

The first three proteins form a core complex that consists of one Toc159, four to five Toc34s, and four Toc75s that form four holes in a disk 13 nanometers across. The whole core complex weighs about 500 kilodaltons. The other two proteins, Toc64 and Toc12, are associated with the core complex but are not part of it.

Toc34 and 33

Toc34 is an integral protein in the outer chloroplast membrane that's anchored into it by its hydrophobic C-terminal tail. Most of the protein, however, including its large guanosine triphosphate (GTP)-binding domain projects out into the stroma.

Toc34's job is to catch some chloroplast preproteins in the cytosol and hand them off to the rest of the TOC complex. When GTP, an energy molecule similar to ATP attaches to Toc34, the protein becomes much more able to bind to many chloroplast preproteins in the cytosol. The chloroplast preprotein's presence causes Toc34 to break GTP into guanosine diphosphate (GDP) and inorganic phosphate. This loss of GTP makes the Toc34 protein release the chloroplast preprotein, handing it off to the next TOC protein. Toc34 then releases the depleted GDP molecule, probably with the help of an unknown GDP exchange factor. A domain of Toc159 might be the exchange factor that carry out the GDP removal. The Toc34 protein can then take up another molecule of GTP and begin the cycle again.

Toc34 can be turned off through phosphorylation. A protein kinase drifting around on the outer chloroplast membrane can use ATP to add a phosphate group to the Toc34 protein, preventing it from being able to receive another GTP molecule, inhibiting the protein's activity. This might provide a way to regulate protein import into chloroplasts.

Arabidopsis thaliana has two homologous proteins, AtToc33 and AtToc34 (The At stands for Arabidopsis thaliana), which are each about 60% identical in amino acid sequence to Toc34 in peas (called psToc34). AtToc33 is the most common in Arabidopsis, and it is the functional analogue of Toc34 because it can be turned off by phosphorylation. AtToc34 on the other hand cannot be phosphorylated.

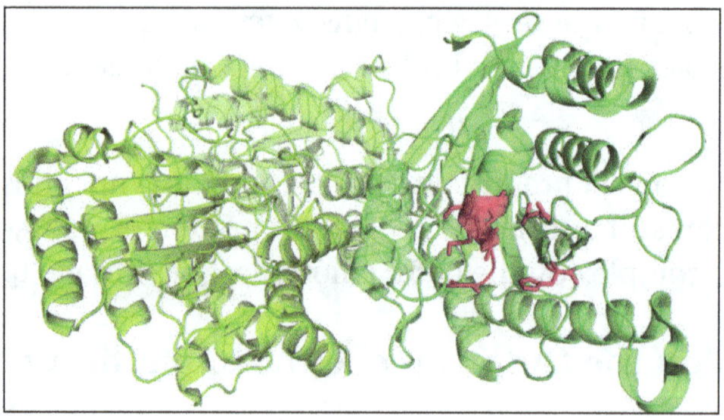

Toc34 from a pea plant. Toc34 has three almost identical molecules (shown in slightly different shades of green), each of which forms a dimer with one of its adjacent molecules. Part of a GDP molecule binding site is highlighted in pink.

Toc159

Toc159 is another GTP binding TOC subunit, like Toc34. Toc159 has three domains. At the N-terminal end is the A-domain, which is rich in acidic amino acids and takes up about half the protein length. The A-domain is often cleaved off, leaving an 86 kilodalton fragment called Toc86. In the middle is its GTP binding domain, which is very similar to the homologous GTP-binding domain in Toc34. At the C-terminal end is the hydrophilic M-domain, which anchors the protein to the outer chloroplast membrane.

Toc159 probably works a lot like Toc34, recognizing proteins in the cytosol using GTP. It can be regulated through phosphorylation, but by a different protein kinase than the one that phosphorylates Toc34. Its M-domain forms part of the tunnel that chloroplast preproteins travel through, and seems to provide the force that pushes preproteins through, using the energy from GTP.

Toc159 is not always found as part of the TOC complex it has also been found dissolved in the cytosol. This suggests that it might act as a shuttle that finds chloroplast preproteins in the cytosol and carries them back to the TOC complex. There isn't a lot of direct evidence for this behavior though.

A family of Toc159 proteins, Toc159, Toc132, Toc120, and Toc90 have been found in Arabidopsis thaliana. They vary in the length of their A-domains, which is completely gone in Toc90. Toc132, Toc120, and Toc90 seem to have specialized functions in importing stuff like nonphotosynthetic preproteins, and can't replace Toc159.

Toc75

Toc75 is the most abundant protein on the outer chloroplast envelope. It is a transmembrane tube that forms most of the TOC pore itself. Toc75 is a β-barrel channel lined by 16 β-pleated sheets. The hole it forms is about 2.5 nanometers wide at the ends, and shrinks to about 1.4–1.6 nanometers in diameter at its narrowest point wide enough to allow partially folded chloroplast preproteins to pass through.

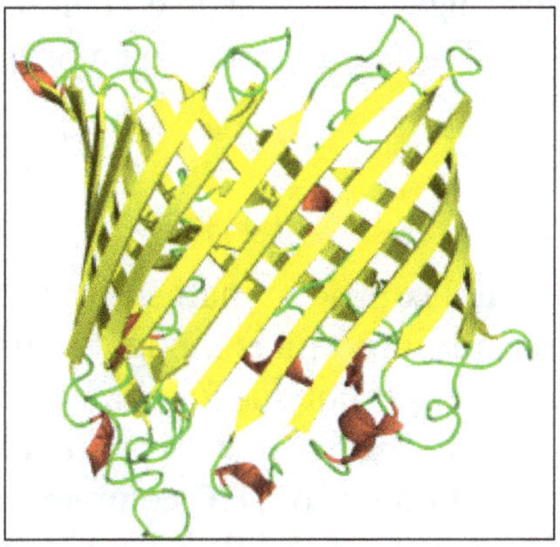

β-barrel The general shape of a β-barrel is a hollow cylinder lined by multiple β-sheets. Note that the protein depicted is not Toc75 specifically.

Toc75 can also bind to chloroplast preproteins, but is a lot worse at this than Toc34 or Toc159.

Arabidopsis thaliana has multiple isoforms of Toc75 that are named by the chromosomal positions of the genes that code for them. AtToc75 III is the most abundant of these.

The Translocon on the Inner Chloroplast Membrane (TIC)

The TIC translocon, or translocon on the inner chloroplast membrane translocon is another protein complex that imports proteins across the inner chloroplast envelope. Chloroplast polypeptide chains probably often travel through the two complexes at the same time, but the TIC complex can also retrieve preproteins lost in the intermembrane space.

Like the TOC translocon, the TIC translocon has a large core complex surrounded by some loosely associated peripheral proteins like Tic110, Tic40, and Tic21. The core complex weighs about one million daltons and contains Tic214, Tic100, Tic56, and Tic20 I, possibly three of each.

Tic20

Tic20 is an integral protein thought to have four transmembrane α-helices. It is found in the 1 million dalton TIC complex. Because it is similar to bacterial amino acid transporters and the mitochondrial import protein Tim17 (translocase on the inner mitochondrial membrane), it has been proposed to be part of the TIC import channel. There is no in vitro evidence for this though. In Arabidopsis thaliana, it is known that for about every five Toc75 proteins in the outer chloroplast membrane, there are two Tic20 I proteins (the main form of Tic20 in Arabidopsis) in the inner chloroplast membrane.

Unlike Tic214, Tic100, or Tic56, Tic20 has homologous relatives in cyanobacteria and nearly all chloroplast lineages, suggesting it evolved before the first chloroplast endosymbiosis. Tic214, Tic100, and Tic56 are unique to chloroplastidan chloroplasts, suggesting that they evolved later.

Tic214

Tic214 is another TIC core complex protein, named because it weighs just under 214 kilodaltons. It is 1786 amino acids long and is thought to have six transmembrane domains on its N-terminal end. Tic214 is notable for being coded for by chloroplast DNA, more specifically the first open reading frame ycf1. Tic214 and Tic20 together probably make up the part of the one million dalton TIC complex that spans the entire membrane. Tic20 is buried inside the complex while Tic214 is exposed on both sides of the inner chloroplast membrane.

Tic100

Tic100 is a nuclear encoded protein that's 871 amino acids long. The 871 amino acids collectively weigh slightly less than 100 thousand daltons, and since the mature protein probably doesn't lose any amino acids when itself imported into the chloroplast (it has no cleavable transit peptide), it was named Tic100. Tic100 is found at the edges of the 1 million dalton complex on the side that faces the chloroplast intermembrane space.

Tic56

Tic56 is also a nuclear encoded protein. The preprotein its gene encodes is 527 amino acids long, weighing close to 62 thousand daltons; the mature form probably undergoes processing that trims it down to something that weighs 56 thousand daltons when it gets imported into the chloroplast. Tic56 is largely embedded inside the 1 million dalton complex.

Tic56 and Tic100 are highly conserved among land plants, but they don't resemble any protein whose function is known. Neither has any transmembrane domains.

CIRCADIAN CLOCK ASSOCIATED 1

The Circadian Clock Associated 1 (CCA1) gene encodes a MYB-related transcription factor involved in the phytochrome induction of a light-harvesting chlorophyll a/b-protein (Lhcb) gene. Expression of the CCA1 gene is transiently induced by phytochrome and oscillates with a circadian rhythm. Constitutive expression of CCA1 protein in transgenic plants abolished the circadian rhythm of several genes with dramatically different phases. These plants also had longer hypocotyls and delayed flowering, developmental processes regulated by light and the circadian clock. Furthermore, the expression of both endogenous CCA1 and the related LHY gene was suppressed. Our results suggest that CCA1 is a part of a feedback loop that is closely associated with the circadian clock in Arabidopsis.

Photoreceptors and circadian clocks are universal mechanisms for sensing and responding to the light environment. The phase of circadian rhythms can be set by light and dark transitions, but these rhythms can persist even in the absence of such external time cues. In addition to regulating daily activities, photoreceptors and circadian clocks are also involved in the seasonal regulation of processes such as flowering, a developmental response to day length called photoperiodism. However, although the importance of circadian rhythms in plant growth and development has long been recognized, and Arabidopsis mutants with altered clock function have been isolated, no clock component has yet been identified in plants.

Studies of the clock mechanisms of Neurospora and Drosophila have demonstrated that in these organisms the central oscillating mechanisms involve a feedback loop whereby transcription is inhibited as a result of the accumulation of the encoded proteins. Clock genes isolated in both organisms, FRQ and WC-2 of Neurospora and PER and TIM of Drosophila, satisfy a set of basic criteria proposed for components of the circadian oscillator. These genes encode nuclear proteins that are predicted to function by regulating transcription. However, only the WC-2 protein has been shown to bind to DNA. The demonstration that WC-2 is a component of the circadian oscillator also supports a central role of the circadian clock in light responses of Neurospora, since the wc-2 mutant is defective in almost all photoresponses. Recently, putative clock component genes have also been identified in mammals.

There is a wide range of processes in plants that show a circadian rhythm. These include movement of organs such as leaves and petals, stomata opening, stem elongation, sensitivity to light of floral induction, metabolic processes such as respiration and photosynthesis, and expression of a large number of different genes and reviewed by. It is believed that the circadian rhythm of gene expression is part of the underlying mechanism for many, if not all, of the rhythms in metabolic and developmental processes.

Most of the circadian-regulated genes studied previously are related to photosynthesis and have a peak level of expression in the morning. A few exceptions include the cold, circadian rhythm RNA binding protein 2 (CCR2/AtGRP7) genes and the Catalase3 (CAT3) gene, which peak in the evening. The rhythmic expression of these genes is thought to reflect outputs from the circadian oscillator. The CCR2/AtGRP7 protein has been shown to be involved in the oscillation of its own transcript and the transcript of a related gene. However, it is not involved in the circadian rhythms of other genes and is therefore believed to be a slave oscillator that is part of an output pathway of the master clock.

There is an intimate relationship between certain photoreceptors and the circadian clock, including the entrainment and resetting of the clock by light, the "gating" of photoreceptor activity by the clock, and the concerted action of photoreceptors and circadian clocks on common cellular and physiological processes. In plants, phototransduction not only sets the phase, but also affects the amplitude and period of circadian rhythms. Members of the phytochrome family of plant photoreceptors, which can exist in two photochemically interconvertible forms and are involved in regulation of plant development and growth, play important roles in regulating clock activities. Many of the genes regulated by phytochrome are also regulated by the circadian clock.

Circadian Clock Associated 1 (CCA1) is a gene that is central to the circadian oscillator of angiosperms. It was first identified in Arabidopsis thaliana in 1993. CCA1 interacts with LHY and TOC1 to form the core of the oscillator system. CCA1 expression peaks at dawn. Loss of CCA1 function leads to a shortened period in the expression of many other genes.

CCA1 was first identified in Arabidopsis thaliana by Elaine Tobin's lab in UCLA in 1993. Tobin's lab was studying promoter fragments that contribute to light regulation of light-harvesting Chlorophyll A/B Binding Protein (LHCB), and noticed DNA-binding activity that had affinity for a specific light-responsive fragment of the LHCB promoter. This DNA-binding activity was designated as CA-1 because the binding is mostly to cytosine and adenine-rich sequences. They found that this binding activity is necessary for phytochrome response. They later found the gene responsible for this activity by screening the Arabidopsis expression library. CCA1 referred to the CA-1 clone, and now designated the gene responsible for this DNA binding activity. The evolution of circadian clocks in land plants is not understood, because circadian rhythms have received little attention in plants other than angiosperms.

Function

CCA1 is part of a negative autoregulatory feedback loop that is coordinated with the daily environmental changes. Repressed in the dark by other proteins, CCA1's expression is activated when light is sensed by the phytochrome in the plant. After translation, the CCA1 protein needs to be phosphorylated by Casein Kinase 2 (CK2). This phosphorylation is necessary for the protein to form a homodimer and to bind to its target promoters. Hyperphosphorylation, due to the overexpression of CK2, will lead to altered circadian rhythms in the mutants where CCA1 showed shorter mRNA circadian oscillation than in wild-type plants. CK2 overexpression is significant in demonstrating CCA1 is part of the clock. The protein motif CCA1 uses to bind to its target DNA sequences is its Myb-like domain. CCA1 only has one Myb domain, whereas other plant and mammalian proteins could have multiple Myb domains. The presence of only one Myb domain in CCA1 shows its importance influence in the circadian clock. LUX is also an important Myb transcription factor that is necessary for CCA/LHY transcription. This can also help account for problems in the repressilator model. CCA1 is also unusual in that it has the ability to bind to asymmetric DNA sequences. CCA1 acts to suppress the expression of the DNA sequence it binds to. The stability of CCA1 protein is not affected by light or dark. It is regulated by its proteasome. Inhibiting proteasome function leads to a circadian rhythm with a longer period.

CCA1 and the Arabidopsis Circadian Oscillator

The Arabidopsis central oscillator contains several proteins that reciprocally repress genes encoding each other to achieve a negative feedback loop necessary to generate circadian rhythms controlling many clock outputs. CCA1 is a key component of this oscillator. Light induces its transcription, and mRNA levels peak at dawn along with Late Elongated Hypocotyl (LHY). CCA1 and LHY associate to inhibit transcription of the Evening Complex (EC) proteins: ELF4, ELF3 and LUX, which suppresses their accumulation until dusk when LHY and CCA1 protein levels are at their lowest. The EC inhibits transcription PRR9 and TOC1 at night. These, along with the remaining PRR proteins PRR7 and PRR5 are involved in suppressing CCA1 and LHY levels, which increase during the night. CCA1 is further involved in maintaining this loop by inhibiting its own expression.

Homologs

Paralogs

LHY (late elongated hypocotyl) also has a Myb domain and functions early in the morning. Both LHY and CCA1 have similar patterns of expression, which could be induced by light. Single loss of function mutants in both genes result in seemingly identical phenotypes. But LHY cannot fully rescue the rhythm when CCA1 is absent, indicating that they may only be partially functionally redundant. Under constant light conditions, cca1 and lhy double loss of function mutants fail to maintain rhythms in clock controlled RNAs.

Orthologs

The circadian oscillator in rice is similar to the Arabidopsis model, and researchers have used this model as a blueprint for understand the rice oscillator. OsLHY in rice serves a similar function as CCA1/LHY and is thus an ortholog of the gene in rice. OsPRR1 in rice is also an ortholog of TOC1. PpCCA1a and PpCCA1b are orthologs of CCA1 and LHy in the moss Physcomitrella patens. They show rhythms with a period of 1 day like their angiosperm homologs in 24-hour light-dark cycles or constant darkness. However these genes show arrhythmicity in constant light conditions, in contrast to CCA1 : LHY.

Mutants

Mutants such as cca1-1 plants, which lack CCA1 protein, show short period phenotypes for the expression of several genes when assayed under constant light conditions. They also have a period 3 hours shorter than that of the wild-type plant, which demonstrates that expression of LHY, its homolog, cannot fully compensate for the loss of the function of CCA1. Plants that have lost function of LHY and CCA1 (lhy;cca1) lost the ability to stably maintain circadian rhythm and other output phenomena. In one study, lhy;cca1 show photoperiod- insensitive early flowering under long- day (16 hours of light/8 hours of dark) conditions and short day (8 hours of light, 16 hours of dark conditions), and arrhythmicity under constant light conditions. However they retain some circadian function in light/dark cycles, showing that Arabidopsis circadian clock is not completely dependent on CCA1 and LHY activity. Plants with non-functioning LHY and CCA1 show a wavy leaf phenotype in constant light conditions. Mutants also have increased vascular pattern complexity in their leaves, with more areoles, branch points and free ends than wild-type Arabidopsis.

LATE ELONGATED HYPOCOTYL

Plants constantly monitor changes in photoperiod or day length to trigger the flowering cycle at the most appropriate time of the year. It is well established that photoperiodic flowering is intimately associated with the circadian clock in Arabidopsis. In support of this notion, many clock-defective mutants exhibit altered photoperiodic sensitivity in inducing flowering. LATE ELONGATED HYPOCOTYL (LHY) and its functional paralogue CIRCADIAN CLOCK ASSOCIATED 1 (CCA1) constitute the core of the circadian clock together with TIMING OF CAB EXPRSSION 1 (TOC1). While it is known that TOC1 contributes to the timing of flowering entirely by modulating the clock function, molecular mechanisms by which LHY and CCA1 regulate flowering time have not been explored.

The Late Elongated Hypocotyl gene (LHY) is an oscillating gene found in plants that functions as part of their circadian clock. LHY encodes components of mutually regulatory negative feedback loops with Circadian Clock Associated 1 (CCA1) in which overexpression of either results in dampening of both of their expression. This negative feedback loop affects the rhythmicity of multiple outputs creating a daytime protein complex. LHY was one of the first genes identified in the plant clock, along with TOC1 and CCA1. LHY and CCA1 have similar patterns of expression, which is capable of being induced by light. Single loss-of-function mutants in both genes result in seemingly identical phenotypes, but LHY cannot fully rescue the rhythm when CCA1 is absent, indicating that they may only be partially functionally redundant. Under constant light conditions, CCA1 and LHY double loss-of-function mutants fail to maintain rhythms in clock-controlled RNAs.

The circadian clock in plants has completely different components to those in the animal, fungus or bacterial clocks. The plant clock does have a conceptual similarity to the animal clock in that it consists of a series of interlocking transcriptional feedback loops. The genes involved in the clock show their peak expression at a fixed time of day. The peak expression of the CCA1 and LHY genes occurs at dawn, and the peak expression of the TOC1 gene occurs roughly at dusk. CCA1/LHY and TOC1 proteins repress the expression of each other's genes. The result is that as CCA1/LHY protein levels start to reduce after dawn, it releases the repression on the TOC1 gene, allowing TOC1 expression and TOC1 protein levels to increase. As TOC1 protein levels increase, it further suppresses the expression of the CCA1 and LHY genes. The opposite of this sequence occurs overnight to re-establish the peak expression of CCA1 and LHY genes at dawn.

CCA1 is generally a more significant component of this oscillator. Light induces its transcription, and mRNA levels peak at dawn along with LHY. CCA1 and LHY associate to inhibit transcription of the Evening Complex (EC) proteins: ELF4, ELF3 and LUX, which suppresses their accumulation until dusk when LHY and CCA1 protein levels are at their lowest. Four primary pseudo-response regulator proteins (PRR9, PRR7, PRR5 and TOC1/PRR1) perform the majority of interactions with other proteins within the circadian oscillator, and another (PRR3) that has limited function. These genes are all paralogs of each other, and all repress the transcription of CCA1 and LHY at various times throughout the day.

Plants that have lost function of LHY and CCA1 lose the ability to stably maintain circadian rhythm and other output phenomena. In one study, such plants showed photoperiod- insensitive early flowering under long- day (16 hours of light/8 hours of dark) conditions and short day (8 hours of light, 16 hours of dark conditions), and arrhythmicity under constant light conditions. However they retain some circadian function in light/dark cycles, showing that Arabidopsis circadian clock is not completely dependent on CCA1 and LHY activity. Plants with non-functioning LHY and CCA1 show a wavy leaf phenotype in constant light conditions. Mutants also have increased vascular pattern complexity in their leaves, with more areoles, branch points and free ends than wild-type Arabidopsis.

The function of LHY was initially demonstrated by a group in the Steve Kay lab, including Andrew Millar.

LUX

Lux Operon

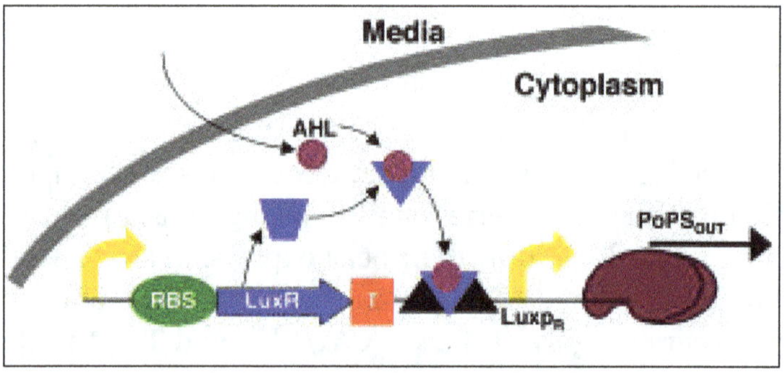

Schematic for the action of the Lux operon.

- Basic Information: The lux operon encodes genes for self-regulation and for the production of luminescent proteins. The most well studied operon is originally isolated from the bacterium Vibrio fischeri, whose operon-produced luciferase produces a yellowish colored bioluminescence of about 490 nm.
- Component Genes:
    - pLuxR: "right hand" lux promoter region.
    - LuxR.
    - LuxI.

- Applications: These are based on the "quorum sensing" abilities of the Lux pathway to acyl-homoserine lactones.

LuxR

LuxR is a constitutively expressed protein that can bind AHL. When bound to AHL it can stimulate transcription from the right hand lux promoter (pLuxR). In the natural system, this promoter controls transcription of the LuxI enzyme leading to a positive feedback loop that increases transcription from the right hand lux promoter. In addition controlling the transcription of luxI, the promoter also controls transcription of luciferase.

LuxI

LuxI is a synthase that converts SAM into a small molecule called an acyl-homoserine lactone (AHL). It can diffuse across cell membranes and is stable in growth media at a range of pH. In the natural system, the pLuxR promoter controls transcription of the LuxI enzyme leading to a positive feedback loop that increases transcription from the right hand lux promoter. In addition controlling the transcription of luxI, the promoter also controls transcription of luciferase.

LUX or Phytoclock1 (PCL1) is a gene that codes for LUX ARRHYTHMO, a protein necessary for circadian rhythms in Arabidopsis thaliana. LUX protein associates with Early Flowering 3 (ELF3) and Early Flowering 4 (ELF4) to form the Evening Complex (EC), a core component of the Arabidopsis repressilator model of the plant circadian clock. The LUX protein functions as a transcription factor that negatively regulates Pseudo-Response Regulator 9 (PRR9), a core gene of the Midday Complex, another component of the Arabidopsis repressilator model. LUX is also associated with circadian control of hypocotyl growth factor genes PHYTOCHROME INTERACTING FACTOR 4 (PIF4) and PHYTOCHROME INTERACTING FACTOR 5 (PIF5).

In 2000, the LUX gene was first sequenced in Arabidopsis thaliana by a team at the Plant Gene Expression Center at UC Berkeley as a part of the Arabidopsis Genome Initiative. In 2003, scientists from the Plant Gene Expression Center and the Genomic Analysis Laboratory at the Salk Institute for Biological Studies collaborated to identify expression of the LUX gene in Arabidopsis using cDNA arrays. In 2005, scientists at the Center for Gene Research at Nagoya University and the Steve Kay lab at the Scripps Research Institute studied null mutations of LUX and the other Evening Complex genes to show that LUX was necessary for circadian rhythms in A. thaliana.

Structure

The LUX gene is located on the third chromosome of Arabidopsis thaliana and contains three exons. Upstream of the LUX gene is a promoter containing a cis-regulatory element known as the "evening element" (EE) with the sequence AAAATATCT. It is overrepresented in evening-expressed genes in the Arabidopsis repressilator. The EE may be bound by Circadian Clock Associated 1 (CCA1) and Late Elongated Hypocotyl (LHY) proteins to suppress expression of LUX. The LUX ARRHYTHMO protein has a length of 323 amino acids and contains a Myb-like GARP family transcription factor DNA-binding domain.

Function

Circadian Oscillator

The LUX ARRHYTHMO protein encoded by the LUX gene participates in the regulation of the Arabidopsis thaliana circadian clock. Along with ELF3 and ELF4, it is a member of the Evening Complex, a component of the Arabidopsis repressilator model of gene regulation. This three-protein complex is expressed and assembled during the evening to repress transcription of the PRR9 gene, which codes for a component of the Midday Complex. LUX likely represses PRR9 via direct binding to a DNA sequence that has not yet been elucidated. PRR9 protein subsequently represses CCA1 and LHY, genes which express components of the Morning Complex. Although LUX and ELF4 are induced by low intensity, non-damaging UV-B radiation, the direct molecular mechanism of light input into the Arabidopsis circadian clock has yet to be elucidated.

Additionally, as a part of the Arabidopsis thaliana repressilator, the LUX gene also represses its own transcription.

Arabidopsis Thaliana Growth and Flowering

The EC binds to promoters of Phytochrome Interacting Factor 4 (PIF4) and Phytochrome Interacting Factor 5 (PIF5), repressing their expression and subsequently inhibiting plant growth in the evening. PIF4 and PIF5 proteins are both basic helix-loop-helix (bHLH) domain transcription factors that are implicated in the induction of Flowering Locus T (FT), which expresses a florigen involved in promoting A. thaliana flowering. Mutants lacking functional LUX are unable to repress PIF4 and PIF5, leading to early accumulation of PIF4 and PIF5 transcription factors and thus premature growth; consequently, LUX mutants often express an elongated hypocotyl phenotype due to excess growth during the night.

Temperature Input

The EC also plays a role in the detection and response to temperature. Despite variations in temperature which would normally reduce the expression of GI (GIGANTEA), LUX, PIF4, PRR7, and PRR9, these genes showed constitutively high expression in LUX (as well as ELF3 and ELF4) mutants. This suggested that LUX mutants abolished the temperature-responsiveness of those clock genes. In addition, ELF3 association to LUX was found to be abolished at high temperatures, suggesting that temperature may play a role in recruiting EC components to their targeted promoters.

Homologs

Paralogs

Paralogs of LUX have been found to act in conjunction with LUX in Arabidopsis circadian clock regulation pathways.

NOX/BOA

In the absence of LUX, ELF3 and ELF4 have also been found to form a complex with LUX paralog NOX, also called BROTHER OF LUX ARRHYTHMO (BOA). NOX is a homologous Myb-like GARP transcription factor that binds to DNA sequences similar to LUX's binding, interacts directly with ELF4, and peaks in the late evening.

Experiments involving artificial microRNA (amiRNA) methods have shown that both NOX and LUX are required to recruit the EC to the PIF4 and PIF5 promoters. There is evidence for NOX having an important role in the regulation of the plant circadian oscillator; overexpression of NOX has been found to have circadian phenotypes of long periods, as well as altered expressions of CCA1, LHY, GI, and TOC1. In particular, overexpression of NOX showed increased amplitudes of CCA1 expression. NOX likely regulates CCA1 through direct binding to the CCA1 promoter, and, conversely, CCA1 protein has been found to bind to the NOX promoter and inhibit NOX expression.

In contrast to LUX, amiRNA knockouts of NOX have shown that NOX is not required for circadian rhythms, suggesting that the functionality of LUX and NOX are not completely redundant. RNAi experiments reducing NOX expression showed a continuation of circadian rhythms, whereas LUX null mutants are arrhythmic. Currently, more research must be done to determine how LUX and NOX differ in their contributions to the EC.

Orthologs

Specific studies of LUX (and ELF3) orthologous mutant alleles have identified variants in flowering and photoperiod-dependent growth.

Sterile Nodes

An ortholog for LUX named STERILE NODES (SN) was discovered in Pisum sativum. The name STERILE NODES came from the observation that photoperiod-responsive P. sativum lines formed more vegetative nodes before flowering compared to less photoperiod-responsive lines. The relationship of LUX and SN as orthologs was concluded based on the discovery of functionally and phenotypically similar mutations in SN and LUX, as well as apparent causal linkages between specific polymorphisms and SN mutant phenotypes. Like LUX, SN was found to be a major gene locus that controls regulation of circadian clock function and photoperiod-sensitive flowering. Also similar to LUX, SN protein is expressed rhythmically when exposed to light-dark cycles.

Gene loci orthologous to Arabidopsis ELF3, ELF4, and GI have also been found in P. sativum, named HIGH RESPONSE TO PHOTOPERIOD (HR), DIE NEUTRALIS (DNE), and LATE BLOOMER1 (LATE1) respectively. However, they have not yet been discovered to form a functional complex equivalent to the EC in Arabidopsis.

HvLUX1

HvLUX1 in Hordeum vulgare has been identified as an ortholog of LUX. The experiment leading to the discovery of HvLUX1 involved a mutation in the early maturity 10 (eam10) locus in the H. vulgare genome. The mutation, called Bowman (eam10), abolished the circadian rhythm observed in H. vulgare flowering. Via high throughput sequencing, HvLUX1 has been identified as a candidate gene for this locus, though its specific mechanism of action in the circadian clock has yet to be demonstrated.

Gene loci homologous to A. thaliana PRR genes (PRR7 and PRR9), ELF3 and FT have also been found in H. vulgare, named PHOTOPERIOD 1 (Ppd-H1), HvELF3, and HvFT1, respectively.

ROC15 and ROC75

Orthologs have been found for all three members of the Arabidopsis thaliana Evening Complex, but it is currently unknown if the EC is formed in species other than A. thaliana. Two orthologs of LUX, ROC15 and ROC75, have been discovered in Chlamydomonas reinhardtii, but orthologs of ELF3 and ELF4 in C. Reinhardtii have not yet been found.

RESISTANCE GENES

In their constant struggle for survival, plants have developed a wide range of defence mechanisms to protect themselves against the attack of pathogens. While some of these resistance strategies rely on simple physical or chemical barriers, more sophisticated biochemical mechanisms based on gene-for-gene interactions between plants and their infectious agents have been reported.

Plant disease resistance genes (R-genes) play a key role in recognizing proteins expressed by specific avirulence (Avr) genes of pathogens. R-genes originate from a phylogenetically ancient form of immunity that is common to plants and animals. However, the rapid evolution of plant immunity systems has led to enormous gene diversification. Although little is known about these agriculturally important genes, some fundamental genomic features have already been described. It has been recently shown that proteins encoded by resistance genes display modular domain structures and require several dynamic interactions between specific domains to perform their function. Some of these domains also seem necessary for proper interaction with Avr proteins and in the formation of signalling complexes that activate an innate immune response which arrests the proliferation of the invading pathogen.

R-genes can be functionally grouped in at least six distinct classes based on the presence of specific domains:

- The CNL class comprises resistance genes encoding proteins with at least a coiled-coil domain, a nucleotide binding site and a leucine-rich repeat (CC-NB-LRR).
- The TNL class includes those with a Toll-interleukin receptor-like domain, a nucleotide binding site and a leucine-rich repeat (TIR-NB-LRR).
- The RLP class, acronym for receptor-like protein, groups those with a receptor serine– threonine kinase-like domain, and an extracellular leucine- rich repeat (ser/thr-LRR).
- The RLK class contains those with a kinase domain, and an extracellular leucine-rich repeat (Kin-LRR).
- The 'Others' class includes all other genes which have been described as conferring resistance through different molecular mechanisms, e.g. Mlo and Asc-1.

Resistance genes (R-Genes) are genes in plant genomes that convey plant disease resistance against pathogens by producing R proteins. The main class of R-genes consists of a nucleotide binding domain (NB) and a leucine rich repeat (LRR) domains and are often referred to as (NB-LRR) R-genes. Generally, the NB domain binds either ATP/ ADP or GTP/GDP. The LRR domain is often involved in protein-protein interactions as well as ligand binding. NB-LRR R-genes can be further subdivided into toll interleukin 1 receptor (TIR-NB-LRR) and coiled-coil (CC-NB-LRR).

Resistance can be conveyed through a number of mechanisms including:

- The R protein interacts directly with an [Avr gene] (Avirulence gene) product of a pathogen (see Gene-for-Gene relationship).
- The R protein guards another protein that detects degradation by an Avr gene (see Guard Hypothesis).
- The R protein may detect a Pathogen-Associated Molecular Pattern or PAMP (alternatively called MAMP for microbe-associated molecular pattern).
- The R protein encodes enzyme that degrades a toxin produced by a pathogen.

Once the R protein has detected the presence of a pathogen, the plant can mount a defence against the pathogen. Because R genes confer resistance against specific pathogens, it is possible to transfer an R gene from one plant to another and make a plant resistant to a particular pathogen.

Many plant resistance proteins are single-pass single-pass transmembrane proteins that belong to receptor kinases and Toll-like receptors.

Humans have been associated with plants since ancient times, and plants in general are an important part for the whole ecosystem on earth, the understanding of plant survival up until now is something that have been an important topic for scientists.

Plants in general are in constant symbiotic relationships with fungus, and even bacteria, but there are cases these symbionts can take advantage of this symbiotic relationship in which case will result in a disadvantageous situation for both organisms. R genes protein syntheses are a way of identifying the pathogen effectors and stop their infection throughout the plant system. Molecules essential for pathogen defense are pattern recognition receptors (PRRs), wall associated kinase (WAKs), receptors with nucleotide-binding domain (NLRs) and leucine-rich repeats (LRRs).All these R proteins play roles in detecting and recognizing pathogen effectors, initiating multiple signal transductions inside the plant cell, these signals transductions will lead to different responses that will aid in pathogen destruction and prevention of further infection. These responses are:

- Production of Reactive Oxygen (ROS).
- Hypersensitive Response.
- Closure of the stomata.
- Production of different chemical compounds (terpenes, phenolic, tannins, alkaloids, phytoalexins).

Note that plants have various mechanisms to prevent and detect pathogenic infections, but factors such as geography, environment, genetic, and timing can affect the recognition pattern of a pathogen or can have an effect on the recognition of avirulent (avr) pathogens in plants.

Pathogen Recognition

R genes synthesize proteins that will aid with the recognition of pathogenic effectors:

Pattern Recognition Receptors

This receptor is often composed of leucine-rich repeats (LRRs). LRRs have a wide range of bacterial (proteins), fungal (carbohydrates) and virulent (nucleic acids) recognition, this means that LRRs recognizes many different molecules but each LRRs usually has a very specific molecule it detects. The ability of PRRs to recognize various pathogenic components relies on a regulatory protein called brassinosteroid insensitive 1 –associated receptor kinase (BAK1). Once the pathogen has been recognized by PRRs the release of a kinase into the nucleus has been transduced triggering a transcriptional reprogramming.

Wall Associated Kinase

The plant cell wall is conformed of pectin and other molecules. Pectin has abundant galacturonic acids which is the compound that WAKs recognizes after a foreign invasion in the plant. Every WAKs (WAK1 and WAK2) has an N-terminal which interacts with pectin in the cell wall when pectin is being degraded to galacturonic acids by fungal enzymes.

Pathogen-associated molecular pattern (PAMPs) and damage-associated molecular pattern (DAMPs) are often identified by lectins which is a protein that binds specific carbohydrates.

Nucleotide-binding Domain and Leucine-rich Repeats

NLRs shift its conformation from ADP state to and ATP state which allows it to send as signal transduction. The activation of NLRs is yet to be completely understood, according to current studies suggest that it is subject to multiple regulators (dimerization or oligomerization, epigenetic and transcriptional regulation, alternative splicing, and proteasome-mediated regulation).

Despite all these differences NLRs, PRRs, WAKs, effector trigger immunity (ETI) and PAMP-triggered immunity (PTI) there are certain similarities such as in the mechanism of signal transduction which includes mitogen-protein kinase (MAPK) cascades through phosphorylation which will be, calcium ion signaling.

An overall overview about the mechanical interaction about a plant defense and the ability of a pathogen to infect a plant would be for instance such a common interaction between bacterial flagellin and receptor-like kinase which triggers a basal immunity sending signals through MAP kinase cascades and transcriptional reprogramming mediated by plant WRKY transcription factors (Stephen T). Also plant resistance proteins recognize bacterial effectors and programs resistance through ETI responses.

Signal Transduction

A plant defense has two different types of immune system, the one that recognizes pathogen/microbes associated molecular patterns (PAMPs), and this is also known as PAMP-triggered immunity (PTI). Plant defense mechanism depends on immune receptors found on the plasma membrane and then the mechanism can sense the pathogen associated molecular patterns (PAMPs) and microbial associated molecular patterns (MAMPs). Detection of PAMPs triggers a physiological change in the cell activated by the pattern recognition receptors (PRRs) initiating a cascade response which through the recognition of PAMPs and MAMPs lead to the plant resistance. The other type of defense is also known as effector-triggered immunity (ETI) which is the second type of defense mediated by R-proteins by detecting photogenic effectors. ETI detects pathogenic factors and initiates a defense response. ETI is a much faster and amplified system than PTI and it develops onto the hypersensitive response (HR) leading the infected host cell to apoptosis. This does not terminate the pathogen cycle, it just slows the cycle down.

Plants have many ways of identifying symbiotic or foreign pathogens; one of these receptors causes fluctuations in the calcium ions and this fluctuation in the calcium ions. A transcription factor plays an important role in defenses against pathogenic invasion.

Pathogen Invasion

Despite the sophiscation of plant defenses, some pathogens have evolved ways to overcome these defenses in order to infect and spread.

Pathogen elicitors are molecules that stimulate any plant defense; among these elicitors we can find two types of pathogen derived elicitors, pathogen/microbe associated molecular pattern (PAMPs/MAMPs), and also there is a second type which is produced by plants known as damage or danger associated molecular patterns (DAMPs). PTI is a way of responding against pathogen actions happening outside the cell, but a much stronger response like ETI is generated in response to effectors molecules. Once there is an induced resistance also known as priming, the plant can react faster and stronger to a pathogen attack. A known priming inducer is called β-aminobutyric acid (BABA) which is a non-protein amino acid. Transgentic plants that produce BABA prime themselves and can effectively defend against Hyaloperonospora arabidopsidis and Plectoesphaerella cucumerina (Baccelli, Ivan). Experiments that put this technique on crops are underway.

Successful pathogens evolve changes in their chemical conformation in order to avoid detection by PRRs and WAKs.

Some viruses have mechanisms that allow them to avoid or suppress the RNA-mediated defense (RMD) that some viruses induce in non-transgenic plants. Further studies has shown that this suppression of the host defense has being done by HC-protease (HCPro) encoded in the Potyviral genome, It was later stablished that HCPro was a mechanism used to suppress post-transcriptional gene slicing (PTGs). Cucumber mosaic virus (CMV) uses different protein called 2b (Pfam PF03263) which is also a suppressor of PTGS in Nicotiana benthamiana.

Even though HcPro and the 2b protein have different protein sequence specific to their own virus, both target the same instrument of defense through different mechanisms.

FLOWERING LOCUS C

Flowering Locus C (FLC) is a major regulator of flowering responses to seasonal environmental factors. Here, we document that FLC also regulates another major life-history transition-seed germination, and that natural variation at the FLC locus and in FLC expression is associated with natural variation in temperature-dependent germination. FLC-mediated germination acts through additional genes in the flowering pathway (FT, SOC1, and AP1) before involving the abscisic acid catabolic pathway (via CYP707A2) and gibberellins biosynthetic pathway (via GA20ox1) in seeds. Also, FLC regulation of germination is largely maternally controlled, with FLC peaking and FT, SOC1, and AP1 levels declining at late stages of seed maturation. High FLC expression during seed maturation is associated with altered expression of hormonal genes (CYP707A2 and GA20ox1) in germinating seeds, indicating that gene expression before the physiological independence of seeds can influence gene expression well after any physical connection between maternal plants and seeds exist. The major role of FLC in temperature-dependent germination reveals a much broader adaptive significance of natural variation in FLC. Therefore, pleiotropy between these major life stages likely influences patterns of natural selection on this important gene, making FLC a promising case for examining how pleiotropy influences adaptive evolution.

One goal toward understanding the genetic basis of adaptation is to accurately interpret geographic patterns of genetic variation in terms of local adaptation. A major challenge in this agenda is to determine the relationship between variable genes and the phenotypes that are exposed to natural selection. A single gene may influence >1 trait, and this pleiotropy may be a considerable constraint on adaptive evolution. The extent to which pleiotropy actually constrains local adaptation depends on how divergent or convergent the complete pathways that regulate the expression of different traits are. Therefore, it is necessary to evaluate the extent of pleiotropy expressed by naturally variable genes, and to characterize the degree of convergence in pathways that regulate the expression of different traits under selection. The gene FLOWERING LOCUS C (FLC), which has previously been shown to be associated primarily with the developmental transition to reproduction, is also strongly associated with the earlier developmental transition of germination.

Among the best characterized genetic pathways regulating a complex life-history trait in plants is the flowering pathway of Arabidopsis thaliana. A central gene in this pathway is FLC, which represses several loci that promote flowering. FLC repression of flowering is relieved by an epigenetic switch that represses the expression of FLC itself via chromatin remodeling. FLC repression can occur after plants experience an extended period of cold (vernalization) or, alternatively, by the autonomous pathway, which senses plant age and ambient temperature. It has been established that FLC expression is promoted by the gene FRIGIDA (FRI), although the exact biochemical basis of the function of FRI remains unknown. As a MADS-box transcription factor (so named for MCM1, AGAMOUS, DEFICIENS, and serum-response factor), FLC directly binds to and inhibits expression of the downstream genes of SOC1 and FT. The repression of these genes, in turn, delays expression of the floral meristem identity genes LEAFY (LFY) and APETALA1 (AP1) to prolong the vegetative phase. Natural allelic variants of FLC are associated with variation in flowering time, either singly or in combination with natural loss of function alleles of

FRI. Variation in these 2 genes has been shown to account for major differences between the winter annual and the rapid-cycling life histories of Arabidopsis. Because FLC has such a large effect on flowering, the maintenance of natural variation in FLC has been thought to be primarily due to geographically variable natural selection on flowering time. Therefore, FLC is unique for ecological studies in that it is one of very few genes for which we have molecular pathway data, natural variation data, and abundant hypotheses concerning how the two are related.

Although much is known about the seasonal and genetic regulation of flowering time, far less is known about the ecological and genetic basis of germination, an earlier life-history developmental transition. However, germination timing is under intense natural selection; it has been shown to account for >70% of the variation in fitness among genotypes in an experimental population of recombinant inbred lines, and it is the life-stage transition that has the largest effect on projected population growth rates in natural populations of A. thaliana. Also, new evidence suggests that the seasonal timing of germination might be a stronger factor in determining the flowering time of Arabidopsis in the field than variation in the genetic basis of flowering time itself. As sessile organisms, the primary way plants determine the seasonal environment of their establishment and subsequent growth is through germination cuing. Therefore, genes associated with natural variation in germination are likely to be under strong natural selection.

Germination and flowering are both fundamental developmental transitions that require precise environmental sensing and responses to multiple seasonal cues to accurately match developmental timing to appropriate seasonal conditions. The combination of these 2 phenological traits determines the overall life cycle and generation time of many plants, including A. thaliana. Because both transitions respond to similar seasonal cues, it is logical to hypothesize that genetic pathways of these 2 life-history transitions share common elements. However, each developmental transition necessarily occurs during different seasons, so some divergence in their regulation is also required. A major regulator of flowering, FLC, controls the germination behavior of seeds in a temperature-dependent fashion, and that natural variation in FLC alleles and FLC expression is associated with natural variation in temperature-dependent germination. FLC regulation of germination is largely maternally controlled and occurs at later stages of seed maturation, and that the signal is transmitted through the downstream pathway involving FT, SOC1, and AP1 to connect with the abscisic acid (ABA) and gibberellins (GA) germination pathways in seeds.

References

- "Ac—Ds (Activator-Dissociator)". Encyclopedia of Genetics, Genomics, Proteomics and Informatics. Netherlands: Springer. 2008. pp. 8–9. doi:10.1007/978-1-4020-6754-9_76. ISBN 9781402067532
- Sakamoto W, Takami T (June 2018). "Chloroplast DNA Dynamics: Copy Number, Quality Control and Degradation". Plant & Cell Physiology. 59 (6): 1120–1127. doi:10.1093/pcp/pcy084. PMID 29860378

- Sandelius AS (2009). The Chloroplast: Interactions with the Environment. Springer. p. 18. ISBN 978-3-540-68696-5
- Functional dissection of circadian clock- and phytochrome-regulated transcription of the Arabidopsis CAB2 gene Proc. Natl. Acad. Sci. USA, 92 (1995), pp. 1500-1504
- Samuel P. Hazen, Thomas F. Schultz, Jose L. Pruneda-Paz, Justin O. Borevitz, Joseph R. Ecker, and Steve A. Kay. LUX ARRHYTHMO encodes a Myb domain protein essential for circadian rhythms. PNAS 2005 102 (29) 10387-10392; published ahead of print July 8, 2005, doi:10.1073/pnas.0503029102
- Troncoso-Ponce MA, Mas P (May 2012). "Newly described components and regulatory mechanisms of circadian clock function in Arabidopsis thaliana". Molecular Plant. 5 (3): 545–53. doi:10.1093/mp/ssr117. PMID 22230762
- Andersen EJ, Ali S, Byamukama E, Yen Y, Nepal MP (July 2018). "Disease Resistance Mechanisms in Plants". Genes. 9 (7): 339. doi:10.3390/genes9070339. PMC 6071103. PMID 29973557

Plant Genomics

CHAPTER 2

Plant genomics deals with the study of genetic compositions, structures, sequencing functions and interactions of an entire plant genome. A few of its aspects are plant genome assembly, microRNA sequencing, genome reference sequencing and resequencing, etc. This chapter will briefly introduce about plant genomics.

Plant genomics aims to sequence, characterize, and study the genetic compositions, structures, organizations, functions, and interactions/networks of an entire plant genome. Its development and advances are tightly interconnected with proteomics, metabolomics, metagenomics, transgenomics, genomic selection, bioinformatics, epigenomics, phenomics, system biology, modern instrumentation, and robotics sciences. Plant genomics has significantly advanced over the past three decades in the land of inexpensive, high-throughput sequencing technologies and fully sequenced over 100 plant genomes. These advances have broad implications in every aspect of plant biology and breeding, powered with novel genomic selection and manipulation tools while generating many grand challenges and tasks ahead.

The Plant Kingdom is a key of the food chain in our planet. Plant domestication by humankind occurred in early societal development, and subsequent agricultural practice and unintentional and intentional plant breeding led to developing productive crop species that provided food and feed products for all living organisms, including humans Plant species are very diverse and there are about 300,000 plant species in the world. Humankind presently grows 2000 plant species in the agriculturally suitable land of 15.5 million square kilometers to fulfill the human diet. Crop domestication with subsequent breeding and farming has created 15 priority crop species, which provide more than 90% of food products. Besides feeding properties, plants supply clothing and housing materials, balance agrobiosenosis and earth ecology, provide medicines and treatment for many diseases, produce energy and biofuels, and have many other key properties and usages to understand life in our planet.

Plant domestication, coupled with the need and desire to get more food and feed products, has resulted in continuous development of breeding and genetics efforts. Early primitive selection attempts have subsequently developed the methods of shuffling traits/characteristics between plant genotypes via controlled sexual crosses that discovered the genetics of key characteristics of crops. Furthermore, the development of biological sciences and understanding of the Mendelian and quantitative genetics of phenotypic variations in plant genotypes, equipped with optimized, targeted, and efficient selection, phenotyping, and statistical methods as well as advanced agrochemical technologies of the past centuries, have revolutionized crop breeding efforts. These advances have resulted in the development of superior crop genotypes that have helped to increase agricultural production. Thanks to the "Green Revolution", the efficient exploitation of plant genetic diversity and plant germplasm resources, novel cultivar development, and better and suitable agrochemical technologies for the past 50 years, the world average cereal crop yield has increased 2.6 times (1.35–3.51), whereas there was 5-fold increase in maize production. There are many such examples of successful conventional breeding efforts. Despite this, food deficiency and human starvation still exist widely and will become even worse with an increase of global human population to ~9 billion by 2050, whereby ~1 billion people may suffer hunger. There is a desire and need to feed the increasing human population, sustain agricultural production, and overcome newly emerging biosecurity issues in the era of global climate change with ever worsening environmental conditions on earth, and societal globalization and technological advances.

These prompted the plant research community to enrich and power the conventional plant breeding and genetics methods with precise tools beyond conventional hybridization, selection, and cultivation/farming practices. This is also dictated by the long duration of conventional breeding and crop improvement, impacted by the limitations in phenotypic evaluations, masking the effect of the environment, polygenic nature of many key traits with many unnoticed minor genetic components, negative genetic correlations between important agronomic traits, linkage drags, and distorted segregation issues in hybridization between diverse genotypes.

To address all these, plant researchers have attempted to decipher the molecular basis of genetic diversities by cloning and sequencing the genes encoding the trait of interest and utilize them in plant breeding as tools in vertical or even via revolutionizing horizontal gene transfers. Progress made toward this goal has elucidated plant genome composition and led to decoding the entire DNA sequences of plant genomes conditioning plant ontogenesis. Here comes "genomics" that was derived from the use of the term "genome" a haploid set of chromosomes coined by Winkeler in 1920. First used in 1986, genomics defined "the enterprise that aimed to map and sequence the entire human genome". Similarly, "plant genomics" is a discipline of plant sciences targeting to decode, characterize, and study the genetic (DNA/RNA) compositions, structures, organizations, and functions as well as molecular genetic interactions/networks of a plant genome Plant genomics aims to develop large-scale high-throughput technologies and efficient tools and methodologies to elucidate the basics of genetic traits/characteristics, genetic diversities, and by-product production; to understand the phenotypic development throughout plant ontogenesis with genetic by environmental interactions; to map important loci throughout the genome; and to accelerate the crop breeding and selection in a genome-wide scale.

Plant genomics research efforts have continuously increased in the past 30 years. The numbers of scientific publications on plant genomics research have drastically increased and reached 17,210 scientific publications in 2015, as indexed in the PubMed database, with its first increase in 2000/2001, following a significant peak after 2010. The first fully sequenced plant genome was the model plant Arabidopsis, which was published in 2000. Since then, almost 50 plant genomes were fully decoded by 2013 and the plant sciences community has finished more than 100 plant genomes by 2015. Furthermore, the plant sciences community extendedly portrayed a sequencing vision of 1001 Arabidopsis accessions and sequencing 1000 plant species that "will have broad implications for areas as diverse as evolutionary sciences, plant breeding and human genetics" while generating many unexpected challenges and grand tasks ahead.

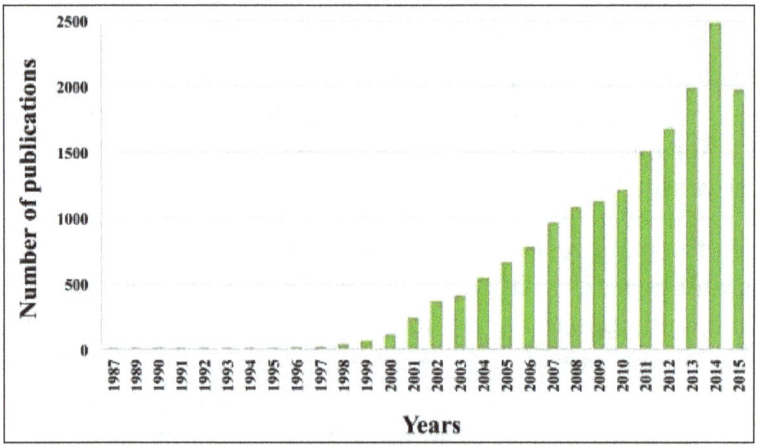

Dynamics of "plant genomics" keyword-retrieved scientific publications in the past three decades.

Genome of Plants and Crop Species

Challenges and Advantages

Compared to other eukaryotic systems, plant genomes are more complex, which create challenges to study its DNA compositions. First of all, the extraction of high-quality DNA from plant tissues, abundantly enriched with phenolic and other metabolic compounds with high affinity to DNA, is conventionally challenging. This interferes with efficient library preparation for whole-genome sequencing, although researchers have optimized methodologies to overcome existing issues.

Furthermore, plant genomes have widely different chromosome numbers, transposon/retro-transposon transcript retention property, and highly varied ploidy levels with many supergenes, pseudogenes, and repetitive elements including low-, medium-, and high-copy number DNA sequences such as transcribed genes, rRNA genes, and retro-elements or short repetitive sequences, respectively. As a result, plant genomes can be 100 times larger in sizes when compared to animal or other model eukaryotic genomes and may contain many paralogous DNA sequences that make sequencing and genome assemblies difficult, which often will generate false-positive errors. For instance, one of the largest examples of sequenced plant genomes, sugarcane (12 Gbs) and hexaploid wheat genome with 17 Gbs in size represents 80% repetitive elements.

Moreover, these massive repetitive "junk" DNA sequences, organized as a simple tandem repeat, repeat single-copy interspersion, inverted repeats, and compound tandem array arrangements, somewhat mask functionally vital single-copy genes, which create a challenge to characterize and clone important individual genes.

Open pollinated, self-pollinated, and clonally propagated plant species have a high level of nucleotide diversity. This can be exemplified by the nucleotide diversity of maize, barley, and grape genomes, where maize genome, for instance, has 10-fold (up to 13%) more polymorphic sites between individual genotypes compared to humans with similar genome size. These polymorphism sites create a challenge in sequence assembly due to the higher rates of nucleotide mismatches to the reference genome.

Lastly, plants tend to have abundant copies of chloroplast genome with two inverted repeat organizations as well as large inversions in some plants with some exchanged regions between nuclear genomes. This creates another challenge in the assembly of repetitive and exchanged regions of chloroplast genomes. The same issue exists in the case of mitochondrial genomes, although it is common for animal genomes as well. All these challenges and complications mentioned above may result in generating fragmented, isolated, and incorrect assemblies in the background of high-copy repeats and paralogous sequences.

However, some specific methodologies and bioinformatics tools have been developed to minimize these challenges. These include the optimized DNA isolation from difficult plant materials use of high-density linkage maps, identification and sorting out of paralogous alleles using local patterns of linkage disequilibrium, and sequencing diploid relatives or ancestor-like genomes of polyploid plants. The use of laser capture microdissection techniques can isolate individual cell types or chromosome or its arm that could minimize the ploidy or paralogy complexities. Moreover, the use of third-generation single-molecule sequencing approaches and novel assembly methods such as optical mapping and long-range Hi-C interactions can also minimize some of challenging cases with the plant genome.

At the same time, along with these challenges and complexities, plants also offer advantages in genome analyses over other eukaryotic systems. This is due to the clonal propagation and indefinite seed storage properties, which create an opportunity for repeated collection of the same DNA samples for sequencing and studying its phenotype multiple times in many generations across replicated environments. There are no ethical issues associated with the multiple uses of plant materials, as it is a sensitive issue for animal cases. The possibility of self-pollination or forced crosses advantageously helps to create highly homozygous samples to reduce existing heterozygosity. There is an opportunity of obtaining double haploid plant genomes. Plant genomes tend to have large chromosomal segments conserved across a large number of taxa in closely related plant species. The collinearity and synteny of plant genomes are very useful to use reference genomes of model species to study homologous and orthologous genes from yet unsequenced genomes.

Sequenced Plant Genomes

The ability to sequence DNA molecules, which was made possible in the 1970s with the introduction of the "plus and minus" sequencing technique of Sanger and Coulson and Maxam and Gilbert, is generally considered to be the starting point of genomics sciences. Later, the simple, long-read chain-terminating dideoxynucleotide DNA sequencing method has become a method of choice to decode genetic sequences. Its eventual automation had extended the capacity and power of this chain-termination sequencing methods to decode the entire genome sequences of living organisms. Because of technological advances and automated sequencing instrumentations, a large-scale sequencing of cDNA libraries made it possible to perform serial analysis of gene expression (SAGE) and expressed sequences tags (ESTs). These were the first genomics technologies for all organisms, including plant genomes. Furthermore, these advances powered by microarray tools routinely used by many individual laboratories worldwide have helped to identify the genome structures and functional and regulatory elements across genomes and have facilitated to develop high-throughput reliable molecular markers for genome/trait mapping studies.

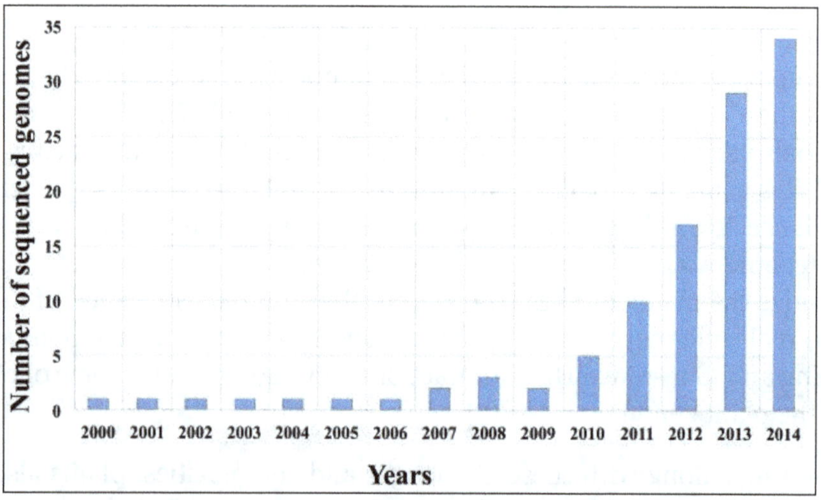

A number of sequenced plant genomes from 2000 to 2014.

The development and generation of massively parallel sequencing technologies provided cost-effective, new-generation sequencing (NGS) platforms that have helped to completely decode the entire genome of many different organisms within a short period. For instance, in plants, the first sequenced genome was a model plant Arabidopsis thaliana with 125 Mbs in size, 25,489 individual genes, and 14% repetitive elements, which was published in 2000. Further, more than 109 plant genomes have been fully sequenced by 2015, including 21 monocots and 83 eudicots, 10 model and 15 non-model plant genomes, five non-flowering plants, and 69 crop species with 6 crop model genomes and 15 wild crop relatives. Following the Arabidopsis model, several rice (Oryza sativa) genomes in 2002 to 2005, black cottonwood (Populus trichocarpa) genome in 2006, and grape (Vitis vinifera) genome in 2008 were fully sequenced. Sequencing whole plant genomes has increased in subsequent years, and 10 plant genomes had been sequenced in 2011. About 80% of sequenced genomes were accomplished in the past 3 years.

The smallest plant genomes sequenced so far are the two eudicot plants: corkscrew (Genlisea aurea) with 64 Mbs genome size and 17,755 genes and bladderwort (Utricularia gibba) with 77 Mbs genome size and 28,500 genes. In contrast, the largest genomes sequenced are from gymnosperm plants, including Norway spruce (19,600 Mbs), white spruce (20,000 Mbs) and loblolly pine (23,200 Mbs). The largest genome sequenced from crop species is the hexaploid wheat (Triticum aestivum) with a genome size of 17,000 Mbs. An average size for all published plant genomes is 1850 Mbs. Per published plant genome data, the gene numbers of the smallest to largest genomes are within the range of 17,755 (corkscrew) to 124,201 (hexaploid wheat) with an average of 40,738 genes for all sequenced genomes. Repetitive elements are highly variable among published genomes that varied from 3% (bladderwort) to 85% (maize, Zea mays) with an average estimate of 46% per genome. These sequenced plant genomes not only provided an updated knowledge on structural compositions and complexities of plant genomes but also elucidated the evolution of gymnosperm and angiosperm plants and specific gene families contributing to the radiation of flowering plants. We learned some direct correlation between genome sizes and gene numbers/repetitive elements, although it does not strictly follow the rules, which was evidenced by several exceptions. For example, one of the largest genomes, Norway spruce, has ~28,000 genes, which is similar to the smallest genome bladderwort. Moreover, medium-sized maize genome (2300 Mbs) or wild tomato (1200 Mbs) contains more or approximately the same (<80%) contents of repetitive elements compared to the largest sequenced genome of loblolly pine (23,200 Mbs).

The results of the first draft of the full genome sequence and assembly of a fresh salad plant leaf chicory (Cichorium intybus subsp. intybus var. foliosum L., 2n=2x=18, and 1.3 Gbs genome size), named as Radicchio in Italian. The results of decoding the full genome of leaf chicory will "extend the current knowledge of the genome organization and gene composition of leaf chicory, which is crucial for developing new tools and diagnostic markers useful for our breeding strategies in Radicchio" and will be an important addendum to the list of sequenced plant genomes.

Sequencing 1001 Genotypes and 1000 Plant Species

The availability of a few whole reference genomes limits our full understanding of ecotypic variations that affect the function and adaptive evolution of plant species in various climatic conditions. It reduces the power of genome-wide tagging of biologically meaningful natural variations. In other words, the general perceptions are that "a single reference genome is not enough" for plant biology to explain and understand the existing natural variations in particular plant species and its populations. It also limits the development of efficient tools for genome analyses. To address this, as mentioned above, the Arabidopsis plant research community has developed a vision of sequencing a larger number of Arabidopsis genotype accessions, including various ecotypic and experimental population samples. As of today, the "1001 genome sequencing project of Arabidopsis accessions" has completed the full genome sequencing of 1100 Arabidopsis accessions "to record the genetic variation in the entire genome of many strains of the reference plant Arabidopsis thaliana" and with the future objective to develop efficient genome analysis tools and software.

To understand the tree of life of the Plant Kingdom and study its evolutionary aspects in comparison to other life forms, the international multi-disciplinary consortium of "The 1000 Plants (oneKP or 1KP) Initiative" has generated a large-scale gene sequencing data for more than 1000 various plant species. Rather than concentrating on single species accessions as in the 1001 Arabidopsis whole-genome sequencing project, the "1KP" project targeted 1000 distinct plant species with the objective of generating only functionally expressed (i.e., transcriptome) gene sequences. The plant species selected for the project had no restriction, and the samples were "chosen to represent every species known to science, across the Plant Kingdom, at some phylogenetically or taxonomically defensible levels". The 1KP sample list consists of 1328 entries broadly grouped by phylogenetically (angiosperm, non-flowering, and green algae species) and by application (agriculture, medicine, biochemistry, and extremophytes). Most of these species have been sequenced for the first time.

Table: Plant species samples chosen for the "1KP" plant genome sequencing project.

Plant Species	*Number of Samples
Phylogenetic Groups	
Angiosperms	830
Angiosperms: Onagraceae samples	50
Non-flowering plants	257
Green algae	241
Application Groups	
Medicinal samples	142
Medicine - Alkaloid samples	30
Medicine - Chemotherapeutic samples	12
Biochemistry - Lipid Biosynthesis samples	15
Agriculture - C3/C4 samples	93
Agriculture - Weeds	25
Extremophyte samples	31
Halophytes samples	18

To date, an average of 2000 Mbs transcriptome sequence data have been generated for these 1KP plant species using 28 Illumina Genome Analyzer next-generation DNA sequencing machines at the Beijing Genomics Institute (BGI-Shenzhen, China). Ultimately, the obtained genomic sequence data will be used to analyze the phylogenetic, taxonomic, and evolutionary relationships of plant species, to study plant speciation, and to determine the timing of gene duplications during speciation events. However, the biggest limitation is associated with sequencing only transcriptomes rather targeting the whole genome, which limits obtaining many non-coding and repetitive portions of genomes. The results of "1001" and "1KP" sequencing efforts will undoubtedly open a new paradigm for plant genomics. The results should not only accelerate crop improvement and boost the agricultural and medicine production worldwide but also help to understand the basics of plant life, evolution, speciation, and plant adaptations to the extreme environments in the era of global climate change and technological advancements.

Crop Improvement in the Genomics and Post-genomics Era

Genomics-assisted Selection or Genomic Selection

At present, the reference genomes for many agricultural plants including specialty crops have been sequenced, as reviewed by Michael and VanBuren, which created a new paradigm for modern crop breeding. Crop breeding, which is powered and enriched by molecular markers, genetic linkage maps, QTL mapping, association mapping, and marker-assisted selection methods in the past century, has now greatly accelerated and become ever productive and efficient in the plant genomics era. This is due to the (1) availability of large-scale transcriptome and whole-genome reference sequences; (2) high-throughput SNP marker collection and cost-effective, automated, and high-throughput genotyping platforms (HTP) and technologies (e.g., genotyping by sequencing or GBS), allowing breeders to screen multiple genotypes within a short time; (3) identification and use of expression QTLs (genetical genomics) in breeding ; and (4) opportunity to perform genome-wide selection (i.e., genomic selection).

The biggest driving force for genomics-assisted crop breeding in the plant genomics era has been the inexpensive sequencing and re-sequencing opportunity for population individuals of genetic crosses and breeding lines. This helps to precisely identify and link genetic variations to the phenotypic expressions, taking into account the rare and private allelic variations that are abundant in crop line population or germplasm resources. Furthermore, the availability of SNP marker collections and automated genotyping platforms provided a better genome converge to perform genome-wide genotype-to-phenotype associations (GWAS). Also, when whole-genome sequences are not available and SNP markers are present in a limited number, the breeders using GBS and HTS platforms can readily genotype their mapping population and can provide genomic selections for the targeted crops of interest. Although it was first applied for animal breeding, recently genomic selection has been successfully applied to a number of plant species including studies using GBS in the context of genomic selection. Most importantly, the application of available genomics tools and a large number of high-throughput DNA markers and new-generation genotyping platforms have made the "breeding by design" possible and have developed "virtual breeding" approaches efficient crop improvement.

The availability of genome sequences and a large number of SNP marker collections also provided the analysis of copy number variations (CNVs) in crop genomes, and their links to the key traits have greatly enhanced the crop improvement programs. Furthermore, although challenges are evident, the opportunity provided by post-genome sequencing advances has help to integrate and enrich genomic selection with key proteome and metabolome markers. This significantly fostered and powered up the breeding of complex traits of crops. Consequently, the knowledge gained through plant genomics coupled with proteomic and metabolomic advances has facilitated the emergence of an innovative approach of "personalized" agriculture through the utilization of chemical genomics. This requires the translation of knowledge and expertise of the pharmaceutical industry on the development of "personalized medicine" to treat each person based on its reaction to the medical drugs into the agriculture. Because of high-throughput genome analysis, it is possible to date that many plant compounds, including herbicides, growth regulators and phytohormones, elicitors, low molecular metabolites (e.g., salicylic acids), and synthetic hybrid chemicals, can be screened for genetic response of individual crop genotypes and to study their mechanism of actions contributing to agricultural productivity. Once identified, highly genotype-specific chemical compounds can be developed that impact better than traditionally applied "fit for all" chemicals/growth stimulators and fertilizers. A combination of such chemical genomics approach, proteomics and metabolomics with genetic engineering, and genomic selection will further provide a way for "personalized" agriculture that sustains crop production.

Novel Transgenomics Tools and Biotech Crops

Crop improvement is also greatly impacted by novel transgenomics and genome editing technologies developed as a result of plant genome characterization and understanding in the era of plant genomics. In the past two decades, a variety of novel transgenomics technologies have been developed to replace or enrich the traditional transgenesis-based genetic engineering and plant molecular biotechnology. These novel technologies include antisense, RNA interference (RNAi), artificial microRNA (amiR), virus-induced gene silencing (VGS), zinc-finger nuclease (ZFN), transcription activator-like effects nucleases (TALENs), oligonucleotide-directed mutagenesis (ODM) of Cibus Rapid Trait Development System (RTDS), and clustered regularly interspaced short palindromic repeats/Cas9 (CRISPR/Cas9) technologies. These novel transgenomics technologies including genome-editing tools,the latter also referred to as genome editing with engineered nucleases (GEEN), are widely developed and utilized to investigate the gene function and apply to solve problems in medicine and agriculture. They are become methods of choice for major functional genomics and biotechnological studies. RNA-mediated genome manipulation (RNAi) tools down-regulate the target genes due to gene silencing effects at transcriptional (TGS) or post-transcriptional (PTGS) levels, whereas GEEN systems help to insert, replace, or remove specific regions of DNA from a genome using artificially engineered nucleases that are referred to as "molecular scissors".

The potential application of RNA-mediated gene silencing methods for crop improvement, including RNAi in plant biotechnology, is huge and the technology has already generated many successful examples in a wide range of technical, food, and horticulture crops. For example, RNAi was used to improve crop yield, food/fiber quality, resistance to pests, and biotic/abiotic stresses, which are being considered for commercialization or are already in commercial production. Employing ODM-mediated single nucleotide editing in Arabidopsis, targeting the BFP gene, has demonstrated a precise edition of CAC to TAC, converting histidine (H66) to tyrosine (Y66) in GFP protein that offered a non-transgenic breeding tool for crops. Similarly, GEEN tools have also provided a new strategy for "trait stacking," whereby several desired traits are physically linked to ensure their co-segregation during the breeding processes. The examples include A. thaliana and Z. mays, where ZFN-assisted gene targeting has helped to heritably insert herbicide-resistant genes (SuRA/SuRB and PAT) into the targeted sites in the genome. Although other GEEN technologies such as TALEN and CRSPR/Cas9 are just picking its application in plants, their utilization in Arabidopsis, maize, rice, potato, wheat, barley and plum holds a great promise and potential for non-transgenic crop genome modification and improvement.

ADVANCES IN GENETICAL GENOMICS OF PLANTS

Ever since the current paradigm of gene transcription preceding biological function, research on gene function has focused on expression studies. With the ever increasing availability of genomic sequences and the introduction of microarray technology, enabling the high-throughput analysis of gene expression, this has rapidly become a favorite tool for many researchers. In a typical microarray experiment specific conditions or developmental stages are studied by comparing expression profiles and determining differences in gene transcription. The object of profiling can be a single genotype showing phenotypic diversity in a spatial and temporal manner, e.g. in different tissues and developmental stages, or when exposed to different growing conditions. In this way, large compendia of expression data have been acquired, providing ontological information of genes involved in developmental control and environmental responses.

Although much of the relationship between the temporal expression of genes and their function can be learned from these analyses, often no information can be obtained about the genetic regulation of transcription or whether expression differences are causal for or a consequence of phenotypic differences. Therefore, instead of comparing different conditions of a single genotype, equivalent samples of different genotypes varying in the trait of interest are often analyzed. These can be natural variants within a species, artificial mutants or transgenics like knockout and over expression lines. Such analyses have shown to be extremely powerful in determining directionality in biological pathways, especially in qualitative traits. However, it becomes extremely difficult to define a proper experimental setup when the trait of interest is complex and has a quantitative character. Geneticists are used to deal with this type of complex traits by using the power of natural variation within species.

In segregating populations, derived from crosses between distinct parents and genotyped with molecular markers, linkage is sought between variation in the trait of interest and genotypic diversity. For this purpose a broad range of software tools and statistical analyses are available. The identified genomic regions explaining the observed phenotypic variation are commonly referred to as quantitative trait loci (QTLs), which can subsequently be used for marker assisted breeding purposes without further knowledge of the underlying genes. Whenever the purpose is to identify the causal genes or even nucleotide polymorphisms (QTNs) underpinning a given QTL, one needs to invest in follow-up analyses for fine mapping and ultimately the cloning of a QTL. This approach however, is very labor intensive and time consuming and confines the classical QTL mapping to a low throughput technique.

Like for many physiological traits, variation in gene expression often shows a quantitative distribution, hence, all the classical statistical tools and concepts for QTL mapping can be applied for its genetic dissection. Thus, subjecting expression variation to linkage analysis identifies genetic regulatory loci, and ideally genes, explaining the observed variation. Knowing the position of genes and their corresponding expression QTLs (eQTLs) renders great opportunities for dissecting quantitative traits. This was first recognized by Jansen and Nap who outlined a concept, coined 'genetical genomics', in which the combination of a genotyped segregating population (i.e. genetics) and genome-wide expression profiling (i.e. genomics) is used to formulate hypothetic regulatory pathways and unravel complex traits in a more high-throughput manner. Analogously, similar approaches can be followed for data derived from other 'omic' technologies such as proteomics (pQTLs) and metabolomics (mQTLs).

The first study reporting a proof of principle of genetical genomics was performed in Saccharomyces cerevisiae. In a relatively small population of 40 haploid segregants from a cross between a laboratory and a wild type strain, it was shown that parental differences in gene expression were highly heritable and amenable to genetic mapping. This first report was quickly followed by more comprehensive eQTL studies in higher eukaryotes and has now been applied in a broad range of taxonomic kingdoms including yeast, nematodes, insects, plants, rodents and humans. All studies demonstrated the power of combining gene expression and genetic analyses to refine molecular pathways involved in complex phenotypes and to identify key driver genes thereof. Moreover, they have shown general and conserved mechanisms of expression regulation which improved our understanding of adaptive strategies and evolutionary concepts.

Genetic Architecture of Gene Expression Variation

The detection of eQTLs depends on a number of factors, which together determine the proportion of genetically regulated genes that can be observed. First, biological factors such as the assayed tissue, developmental stage or environmental conditions and the genotypic diversity present in the mapping population determine which genes are expressed and exhibit allelic variants, respectively. Second, statistical issues like population type and size, genetic map quality, measurement accuracy and the number of genes analyzed determine mapping power and detection thresholds. Because all these aspects vary between different experiments, reported fractions of regulated genes range from only a handful to over 50% of the total gene content.

Given the prerequisite of allelic variation, there can be many reasons why genes are differentially expressed in genotypically diverse individuals of a species. Well-known phenomena are allelic variants of transcription factors and other regulators, cis-elemental variation in promoter sequences, differences in mRNA stability, copy number variation and genomic rearrangements such as translocations, insertions and deletions. The latter include gene loss and duplication, resulting in neo and sub-functionalization. Most of these variations in DNA structure will result in eQTLs but depending on the position of the causal polymorphism, an important dissection is made in local and distant eQTLs. Local eQTLs can be the result of closely linked trans-acting factors but in the majority of cases result from cis-regulatory variation in the genes under study. By definition eQTLs acting in cis affect transcription initiation, rate and transcript stability in an allele-specific manner. In addition, cis-regulated genes might encode regulators affecting the expression of downstream target genes in trans. Although the exact proportion varies between studies the occurrence of cis-acting eQTLs is substantial ranging from one-third to half of the total number of eQTLs.

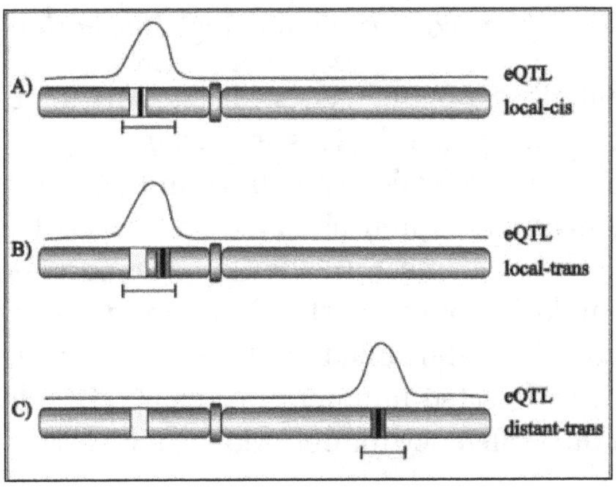

Classification of eQTLs (solid line) based on the expression of the gene under study (light grey box) and location of the causal polymorphism (black bar).

- Local cis-eQTL, result from allelic variation of the gene under study.
- Local trans-eQTL, causal polymorphism within the eQTL confidence interval but not inside the gene under study (no allele specific expression).
- Distant trans-eQTL, gene under study is located outside the confidence interval of its eQTL.

However, because of limitations in mapping resolution, eQTL support intervals may still contain multiple genes and as a result the classification of cis-eQTLs should be used with care. To discriminate true cis-regulatory polymorphisms from local trans-regulation, allele specific expression (ASE) assays can be performed. In such assays a transcribed polymorphism is used to enable discrimination between the parental transcripts and test for allele specific expression in an F_1 hybrid. Because both parental alleles share the same genetic background in F_1 hybrids, and therefore are equally exposed to trans-acting factors, any difference in expression can only be explained by true cis-acting variation. Usually, ASE-assays are performed by single gene qRT-PCR approaches but the recent development of whole genome SNP-tile microarrays (e.g. in Arabidopsis) enables the simultaneous testing of genome-wide ASE.

Although expression differences are treated as quantitative traits in mapping approaches, qualitative differences, characterized by a total lack of expression for one of the allelic variants, can also be observed. The variation in a measurable detection signal can be due to differences in hybridization efficiency, which can be confirmed with genomic DNA hybridization, or genuine loss of transcription. Hybridization efficiency differences are often caused by polymorphisms in the complementary sequences of the microarray probes or mRNA splice variation and are not necessarily accompanied by transcription differences. True transcription variation however, can be caused by strong polymorphisms in promoter regions, premature stop mutations and even the complete absence of genes in one of the parental lines. Both hybridization and true transcription variation will lead to strong cis-eQTLs which can subsequently be used as molecular markers, allowing the construction of high-resolution maps.

The majority of differentially expressed genes will show a quantitative expression profile with complex inheritance patterns. This is because in general genes are regulated by many independent factors which can show up as trans-eQTLs. Because of the multiplicity of regulators and the often-observed epistasis between them, each trans-eQTL can have a relatively small effect. In addition, compared to the direct regulation of cis-eQTLs, the accumulation of stochastic variation in the expression of trans-regulated genes is indirectly also determined by the expression variation of one or more regulators. As a result the detected number of trans-eQTLs relative to the number of cis-eQTLs drops when the stringency for detection is increased.

Whereas cis-eQTLs are inherently associated with the gene in which they reside, a single gene can be responsible for the appearance of multiple trans-eQTLs throughout the genome. As a consequence the genome-wide distribution of cis-eQTLs is dependent on local gene density, although variation in chromatin structure can have an impact on the exposure of eQTLs. The distribution of trans-eQTLs however, can deviate substantially from what can be expected based on gene density. The identification of so-called hot spots, genomic regions with a high density of trans-eQTLs, can be explained by major regulators, e.g. transcription factors, which influence the expression of many downstream genes. In Arabidopsis this was illustrated by the large number of genes mapping to the ERECTA locus, a gene well-known for its pleiotropic effects on many morphological and developmental traits. These findings suggest that the effects of key-regulators in gene expression are progressed to the phenotypic level. This was recently confirmed in a QTL study comparing transcript, protein and metabolite data with phenotypic traits.

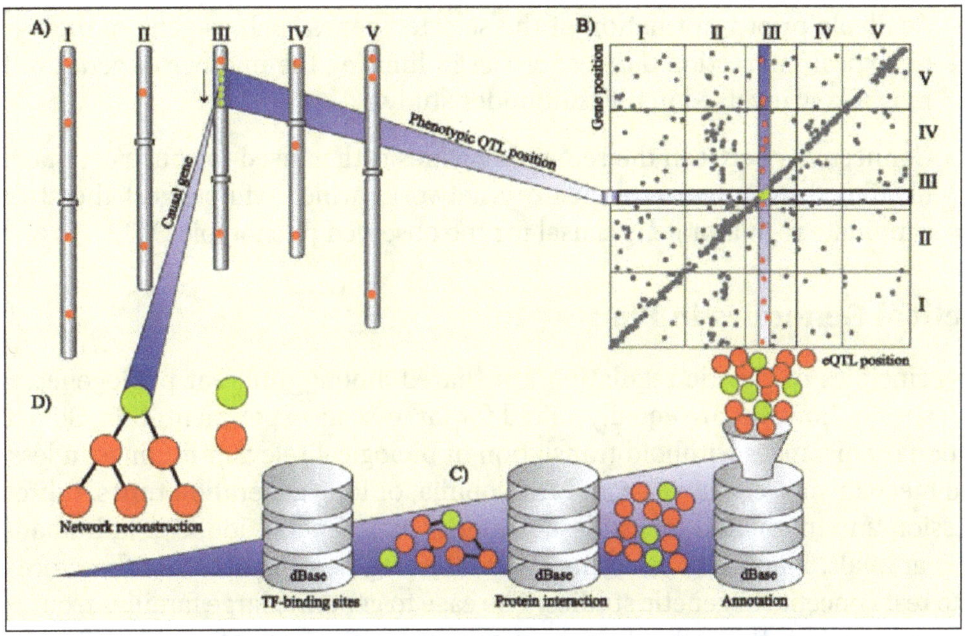

Candidate gene selection with a genetical genomics approach.

Here, only a limited number of QTL hot spots with major, system-wide effects were detected, indicating that most of the genotypic variation is phenotypically buffered. These findings support the theory of biological robustness where hotspots indicate fragilities in this genetic buffering system. Until now only a few reported hotspots have been verified and the number of detected hotspots is far from consistent between different genetical genomics studies. The latter reflects differences in the analyzed populations; species and conditions used and additionally might be the consequence of different statistical procedures used to identify eQTLs. Because of the difficulties in cloning QTLs and the large biological relevance of hotspots, additional sources of information are often used to reduce the number of candidate genes or even predict the causal regulator. Such methods use information on gene ontology, (co-)expression, transcription factor binding sites and targets, ChIP-Seq and protein-protein interaction. Together with computational methods such as regulatory modeling this can severely reduce the number of candidate genes and prioritize remaining candidates for further experimentation.

- Five chromosomes of Arabidopsis thaliana (I-V) with a QTL support interval for a phenotypic trait indicated on chromosome III.
- eQTL plot showing the position of genes and their corresponding eQTLs. Genes within the support interval can be causal for the observed QTL, of which cis-regulated genes, indicated in light grey, represent the strongest candidates. Genes outside the QTL support interval but regulated in trans by the same locus, indicated in dark grey, might be involved in the biological process under study and represent downstream effects of the QTL.
- Available prior information of the selected genes such as gene onthology and biological interaction data can assist in limiting the number of genes to those most likely involved in the trait under study.
- Connectivity between the remaining genes is then used to construct maximum likelihood hypothetical regulatory networks which will suggest the strongest candidate regulator gene causal for the observed phenotypic QTL.

Genetical Genomics in Plants

Many principles of genetic regulation are shared among different phylogenetic taxa. Not all species however are equally suited for large-scale experimentation. Sometimes evolutionary distances withhold translation of biological relevant findings in less conserved mechanisms, e.g. in yeast and Drosophila, or long generation times, inbreeding depression and moral and ethical issues hinder experimentation, e.g. in humans and other mammals. Plants, representing one of the largest kingdoms, are therefore often used to test concepts in genetic studies. The ease to generate large families from experimental crosses and the ability to store genotypes in the form of seeds or clonal propagation make plants ideal subjects to study the mechanistic basis of genetic regulation of traits.

Arabidopsis as a Reference Plant

The comprehensive resources which are available for Arabidopsis thaliana, such as a whole genome sequence, a large collection of natural variants and an ever-increasing number of molecular tools, made it the favorable model for genetical genomics research. As a non-obligate selfing species Arabidopsis combines the ability to cross-pollinate with high tolerance to inbreeding. Together with its short generation time and high reproductive success rate this enables the fast generation of large experimental populations such as Recombinant Inbred (RI) and Introgression Line (IL) populations. The availability and immortal character of such populations enable the accurate estimation of phenotypic values through replicated measurements and allows the testing of traits in different environments.

Traditionally QTL studies of 'classical' physiological traits in RIL populations are followed by mendelizing detected QTLs in near isogenic lines (NILs) for detailed analyses. By isolating QTLs from their genetic background it becomes much simpler to study their genetic effect and relate resulting phenotypes to other processes. Because it is expected that much of the phenotypic variation is the resultant of differences in gene expression and phenotypic perturbation in turn leads to transcriptional reprogramming, data mining for relationships between trait values and expression levels has become a common tool. Very often mutants, knockouts or over-expression lines are used for these purposes, in which the effect of a single gene perturbation is tested on both the phenotypic and the expression level. For complex traits however, the causal genes leading to altered phenotypes are often not known and QTL analyses only identify genomic regions containing such genes. Nevertheless, using RIL populations to identify QTLs for a phenotypic trait and subsequently analyzing NILs for expression differences can be a powerful alternative to explore the functional relationship between genotype and phenotype. Although the regions spanned by NILs can still contain hundreds of genes, of which many may display allelic variation between accessions, the cis-regulated genes are strong candidates explaining phenotypic diversity. High detection stringency can limit the number of differentially expressed genes to a reasonable number of candidate genes with strong local eQTLs.

The availability of a whole genome sequence in Arabidopsis provides unique opportunities, especially when multiple (epistatic) phenotypic QTLs are detected. Knowing the position of genes allows the identification of strong cis-regulated genes collocating with phenotypic QTLs. An early eQTL study in Arabidopsis analyzed genome-wide gene expression in a limited population of only 30 individuals, mimicking shoot regeneration conditions. Two of the eQTL hotspots found coincided with shoot regeneration QTLs. The most significant eQTLs within these hotspot regions showed local chromosal linkage with their corresponding genes but the majority acted distantly. These results suggest that heritable cis-regulated expression changes of key-regulators determine in trans the expression of many genes related to differences in shoot regeneration efficiency between accessions. It also indicates that a long signaling cascade may exist between the causal genotypic polymorphism and the eventual phenotype.

In contrast to the former study it is not always necessary to combine phenotypic measurements with expression analysis. Often, many genes are known to play a role in the exposure of certain traits without knowledge about the genetic regulation of these genes. Specific analysis of such genes can help to identify common regulators. In the first genome-wide eQTL study in Arabidopsis, using a complete RIL population (162 lines), this concept was used to predict possible key-regulators of flowering time and circadian rhythms. The benefits of using large populations for eQTL studies became also apparent in another study where expression analyses were performed in a RIL population of 211 individuals. Whereas in the majority of cases only a single QTL could be detected per differentially expressed gene in the aforementioned studies, here the expression of many genes was controlled by multiple eQTLs. Moreover, a much larger fraction of genetically regulated genes was identified with a higher proportion of trans-regulated genes of which the vast majority exhibited small effects.

The studies performed in Arabidopsis show that the statistical power to detect eQTLs depends largely on population size. Nonetheless, it cannot be excluded that differences in the analyzed tissues, developmental stages and populations used, such as parental variation, linkage distortion and recombination frequency, are responsible for part of the observed differences. All studies however, clearly demonstrated that variation in gene expression is for a large part genetically controlled, with much stronger effects of cis-eQTLs compared to trans-eQTLs. In general, cis-eQTLs also exhibit much higher heritability values and are obvious candidates to act as causal regulators of genes showing trans-eQTLs in the hotspots that could be detected. The detection of regulatory loci for gene expression and the elucidation of their interaction networks might therefore provide the research community with a powerful tool to unravel the complex nature of natural variation in quantitative traits.

Applications in Crop Species

Genetical genomics studies in Arabidopsis and other model species have shown the enormous benefits of the availability of an annotated genome sequence. However, until now full annotated genome sequence information for agronomical important species is only available for a limited number of species, including Oryza sativa, Populus trichocarpa,Vitis vinifera and papaya. This relatively low number of sequenced crop species can be explained by their often immense (polyploid) genome sizes and the highly repetitive nature of many crop genomes. Nevertheless, sequence efforts for many more species, are ongoing and the increasing power of next-generation sequencing will soon lead to an almost unrestricted availability of genomic sequence information. Although an annotated genome is a valuable resource for the comparison of the genomic position of genes and their respective eQTLs, for most crop species this is not feasible yet. Nonetheless, several studies in crops for which genetic maps are available have shown that comprehensive genetical genomics approaches are possible without the need for annotated genome sequences.

Illustratively, one of the first large genetical genomics experiments was performed in an economically important species, viz. Eucalyptus. QTL analysis of transcript levels of lignin-related genes showed that their mRNA abundance is regulated by two genetic loci coinciding with QTLs for stem diameter growth. Genetic mapping of some of the candidate genes showed that most of the lignin genes are under control of a trans eQTL hotspot which suggests that transcription of many of the genes in this pathway are under a higher level of coordinated control. A strong cis-regulated gene encoding S-adenosylmethionine synthase, collocating with the growth and transcription QTLs, was presented as the possible rate limiting step in lignin biosynthesis and as such a strong candidate for the observed QTLs.

In some crops the required availability of genomic sequence data for large-scale classification of cis/trans eQTLs can be circumvented by making use of synteny with other species. In wheat, synteny with rice was used to assist the physical mapping of wheat genes. A genetical genomics approach was conducted in a segregating population of 41 doubled haploid (DH) lines to study agronomic important seed quality parameters. Assuming that the most significantly different expressed genes were cis-regulated, a selection of genes was subjected to synteny analyses. This enabled the positioning of genes with biological relevant linkage to phenotypic traits in a species for which full genome sequence is not available yet.

In the absence of genome-wide micro-arrays, expressed sequence tag (EST) libraries allow the construction of species specific sub genome-scale microarrays. In maize, cell-wall digestibility, which is the major target for improving the feeding value of forage maize, was analyzed in a RIL population. In addition forty extreme RIL lines were hybridized on a small microarray with 439 preselected candidate ESTs for cell-wall digestibility genes for which 89 eQTLs could be mapped. One eQTL hotspot collocated with a cell-wall digestibility related QTL. The application of genetical genomics approaches can be of special interest here when the detection of eQTLs is combined with ASE assays. The thus identified cis-regulated genes can then be positioned on the genetic map where they may serve as candidate genes underpinning phenotypic QTLs.

An interesting alternative for species for which no (EST) sequence information is available at all, and hence no microarrays can be produced, is a gel-based cDNA-AFLP approach. Here AFLP band intensities, reflecting expression differences, are profiled for a large proportion of the transcribed gene pool enabling standard eQTL analyses procedures. AFLP bands showing significant eQTLs can subsequently be sequenced to obtain the identity of the gene from which the fragment derived. Additionally, the cDNA-AFLPs can be used to construct a genetic map.

The examples given above show that genetical genomics is not necessarily restricted to model species but can be applied to any species in which experimental crosses are possible even in the absence of genomic sequence or genetic map information. The potential of combining phenotypic QTL analysis with gene expression traits is shown in a number of economically important species, e.g. Populus, cotton, rice and sunflower. The application of genetical genomics is particulary promising in breeding programs of crops that take advantage of hybrid vigour. The eQTLs involved in heterosis will segregate consistently in a F_1 backcross population thereby identifying valuable targets for marker assisted breeding for the best combination of alleles in the parents of the hybrid.

Network Reconstruction

Genetical genomics harbors the potential to dissect the genetic regulation of a specific biological process. Therefore, methods to reconstruct regulatory networks from eQTL data have obtained much attention. Prioritizing on cis-eQTLs that collocate with a phenotypic QTL is a valuable approach for causal gene discovery, but in many cases little is known about the global regulation, interaction and function of genes that control a biological process. Identification of a set of genes with a trans-eQTL at an identical position can help to dissect genetic variation that is influencing an entire pathway and can lead to the identification of initiating polymorphisms upstream in a network. Questions about the regulatory level at which trans polymorphisms act in the global gene expression network and what their effect is on phenotypic variation and heritability can only be addressed when eQTLs are further dissected.

With a genetical genomics approach one can use the natural genetic variation as a source of perturbations to elucidate the structure of networks. In a summation approach eQTLs for all genes in the analysis are simply superimposed to identify common regions which control many genes. Such an approach does not require any a priori network information but applies subsequent Gene Set Enrichment Analysis (GSEA) using gene onthology (GO) annotation or other descriptors to test whether selected genes share a common biological function. If the network under study is largely known or at least predicted, an a priori analysis can be performed. Here, the expression levels of individual genes in the network are converted into a common measure for the expression level of the entire network which is then used as the trait for QTL analysis. This strategy was tested in an Arabidopsis RIL population for 20 gene expression networks and resulted in statistically significant network variation for eighteen of the 20 predefined networks. Combining summation, GSEA and a priori network analyses allows the generation of a more specific hypothesis about phenotypic effects of network eQTLs. In a study using 175 genes, selected to be involved in regulation of flowering and circadian rhythms, 83 genes showed an eQTL. By combining co-expression analysis, which becomes feasible for microarray compendia of large populations, and positional information of genes and their eQTLs, it was possible to construct regulatory networks of key-regulators and their target genes, predicting unknown relationships and confirming common knowledge.

Pre-selection of known pathways can obviously hinder the elucidation of novel networks in a species, for which much effort is made to develop methods to translate eQTL data into network information using an a posteriori approach. As the precise balance of active components within a tightly controlled biological pathway is in part maintained by coordinately regulated gene expression, this creates possibilities to model networks by exploring co-expression of untargeted genes. To validate this hypothesis, gene expression in liver from a population of 60 mice with variation in diabetes susceptibility was analyzed. The combination of correlation analysis across a genetic dimension and linkage mapping enabled the identification of regulatory networks, functional predictions for uncharacterized genes and characterization of novel members of known pathways. A similar approach in Drosophila, complemented with information about gene ontology, tissue specific expression and transcription factor binding sites, led to the construction of multiple interconnected networks with biological relevance for phenotypic traits.

Understanding the mechanisms underlying trait regulation requires the identification of specific causal polymorphisms. For this purpose sophisticated self-learning algorithms have been developed which make use of conservation, type and position of a particular SNP to prioritize causal regulators by estimating the likeliness that it plays a causal role in gene expression variation. Extending such approaches might also provide the means to distinguish whether variation in gene expression or a regulatory network is the cause or a consequence of an altered phenotype, resulting in the construction of probabilistic directional networks. Defining such causal networks is also known as reverse engineering, because it aims at understanding how the system works as an integrated whole instead of only defining the functionally related components.

Next Level Networks: Integration of other Omics Data

Although phenotypic variation can be partly explained by genetic variation in gene expression, this alone does not fully cover the possible differences in the regulatory mechanisms of an organism. Similar transcript levels of allelic gene variants can still result in varying protein levels because of variance in translational activity, protein degradation and post-translational modifications. Furthermore, variation in coding sequences can alter protein function resulting in a flexible metabolome in terms of chemical structure and function. Integrating 'omics' data such as gene expression, SNPs, metabolomics and proteomics in genetic studies can therefore reduce the number of candidate genes for a given QTL from hundreds to a manageable list without excluding regulatory mechanisms a priori. Because of the analytical complexity in analyzing large numbers of protein samples, genetical proteomics studies are limited but advances made in biochemical detection have already enabled the large-scale untargeted genetic analysis of metabolic content.

The complex relationship between different levels of regulation was illustrated in a study integrating parallel QTL analyses of the expression of genes, activity of encoded enzymes and metabolites involved in primary carbohydrate metabolism. It could be shown that regulation acted on each of the intermediary levels of the path from genotype to phenotype. Although seemingly specific independent regulation could be observed for each analyzed trait, a strong interconnectivity existed between them resulting in coherent systematic differences between population individuals.

The importance of the tight regulation of such an essential component in plant development as primary metabolism was also demonstrated in an Arabidopsis RIL population where plant biomass was related to the metabolic profile. Again, no relationship could be observed between individual metabolites and plant growth but a strong canonical correlation was observed between biomass and a specific combination of metabolites in central metabolism. The power of large-scale metabolomic profiling combined with detailed morphological analysis was also shown in tomato. Significant QTLs could be detected for the accumulation of a large number of primary metabolites together with loci that modify yield-associated traits. With this information a correlation network revealing associations between phenotype, metabolic content and nutritional value could be generated. These studies show that analyzing phenotypic traits and metabolic profiles in a genetic mapping population has great potential for the generation of biomarkers in breeding programs.

Whereas primary metabolites are essential in central metabolism governing growth and development, plants also accumulate large amounts of secondary metabolites. These are believed to be less essential but may play an important role in the adaptation of plants to local environments. Since Arabidopsis can be found in a wide variety of habitats, variation in secondary metabolism might explain much of the evolutionary success of the species. A large untargeted screen of variation in secondary metabolic composition indeed revealed a high proportion of genetically controlled compounds. The highly flexible nature of the metabolome was clearly shown by the fact that more than one-third of the compounds present in the RILs were not detected in either parent but was the result of recombination in biosynthesis pathways. The genetic information obtained from such studies is of great value for the construction of molecular biosynthesis networks, especially if they can be combined with expression data.

This strategy was applied in the genetic analysis of glucosinolate biosynthetic networks which were studied at both the transcriptional and metabolic level. In all cases, variation in gene expression also affected the accumulation of metabolites but epistasis was detected more frequently for metabolic traits as compared to transcript traits. Within such an a priori defined framework it was possible to identify and unravel complex regulatory mechanisms like metabolic feedback loops in which metabolic content regulated gene expression and vice versa.

Plant genomes vary in size, ploidy, and chromosome number. Sizes can range from 120 Mb for Arabidopsis thaliana to 50,000 Mb for some lilies.

Genomics has emerged as a science of its own, being much more than the sum of genome-wide methods. This is particularly the case when considering plant genome studies. The wealth of data that is becoming available on trait-conferring genes, on the number of gene copies, and soon on hotspots for recombination, is of high importance to plant breeding.

The first genome sequencing of a higher plant that of Arabidopsis thaliana, a common weed of the Brassicaceae family, has now reached the halfway stage. This progress has been greatly stimulated by other genome projects, from human to yeast. Meanwhile, tools are being developed to analyze the wealth of information that becomes available once an entire genome has been sequenced. Indeed the structure and function of genes in a genomic sequence can still only be predicted with great difficulty. Thus, there remains a huge task for functional genomics to assess gene function on a genome-wide level.

Structural Genomics

Mapping and Sequencing

The basis for genome analysis in plants is often a genomic map, which can be either a genetic map based on information from both visible and molecular markers, or a physical map, in which yeast artificial clones (YACs) and bacterial artificial clones (BACs) are aligned with the chromosomes. Such maps are available for a wide range of plants, although only a genetic map is available for a limited number of agriculturally important crops, because the genome has not yet been cloned into YAC or BAC libraries.

In addition, the sequencing of plant genomes has been initiated. Because knowledge in plant molecular biology is lagging somewhat behind the yeast and animal field, the prime interests are the transcribed regions of the genome. Therefore, sequencing projects were initiated with expressed sequence tags (ESTs), which are single sequence reads on randomly selected cDNA clones.

EST Sequencing

For both Arabidopsis and rice, large EST collections of approximately 35 000 clones are already available, as well as some small EST collections of approximately 5000 clones, for instance for poplar and soybean. In the dbEST section of GenBank the number of ESTs available for other plants can be followed. For the time being, in contrast to genomic sequencing, the growth in base pairs is rather slow, but there is a remarkable expansion in the number of new (crop) species for which an EST collection is available.

Genome Sequencing

Arabidopsis was chosen for the first plant genome sequencing project, mostly because of the small size of its genome (120 Mb), and the fact that it has become the model plant for a wide range of studies in plant sciences. The first analysis of the Arabidopsis data has shown that this genome is extremely gene-rich and poor in repeated elements. On average, there is one gene every 4–5 kb, the average number of introns is five, with a mean size of approximately 160 bp. The total gene content of Arabidopsis is estimated to be approximately 20 000–25 000. About half of the predicted genes can be assigned to a functional category based on their similarity to known proteins or motifs. In addition, the genome sequences have taught us that even in a relatively small genome, such as Arabidopsis, evidence for ancient genome duplications can be found.

The rice genome sequencing program is also taking shape. Rice has been selected as the second model plant because it is a monocotyledonous species with a small genome size (roughly 4-fold that of Arabidopsis) and because of its importance as a crop. An international team is setting the primary guidelines and the first data are currently being released.

More plant genomes will be sequenced in the following years, probably at low coverage, with the coming of a new generation of high throughput sequencers, some based on the principle of mass spectrometry. To be considered are a nitrogen-fixing legume, such as Medicago sativa, and trees, such as eucalyptus (450 Mb). Also the US National Science Foundation is funding a mapping and cloning project that sets the stage for a large genome sequencing project of corn. The focus will be on the regions in which valuable information is probably present rather than on sequencing the whole genome (3 billion bases full of repetitive DNA). When several plant genomes will have been sequenced in the coming years insights into species evolution will be gained from comparing genes and genome organization.

Genome Organization

Recent studies indicate common aspects in the characteristics of many plant genomes, including in the structure of chromosomal components, such as telomeres and centromeres. However, plant genomes differ in their gene organization. Barakat et al. analyzed the genome of Arabidopsis and have shown that its organization is different from that of the genomes of Gramineae. Genes in Arabidopsis are fairly evenly distributed, whereas studies of maize, rice, and barley have shown that the vast majority of genes are clustered in long DNA stretches (so-called gene space) that are separated by expanses of gene-empty DNA. More than 50% of the maize genome is composed of interspersed repetitive DNAs, primarily retrotransposons that insert between genes.

Mobile DNAs in plants can be grouped in two classes: those that transpose as DNA molecules and those that transpose as an RNA intermediate. The better-known elements such as Ac and En/Spm fall into the first class. Their copy number is usually rather low, with the exception of miniature inverted-repeat transposable elements (MITEs) that can be found at a few thousand copies per genome. The elements of the second class are usually referred to as retroelements. In large plant genomes, such as maize and barley, these elements, which contain long terminal repeats (LTRs), range from a few kb to 15 kb in size and make up the majority of the genome. These retroelements are less abundant in plants with a smaller genome.

Comparative Genomics

Comparative genomics is usually based on genomic maps and sequences and studies primarily the more complex genomes, such as those of grasses. Comparative genomics offers possibilities to link plant families through their genomes. These studies will provide keys to understanding how genes and genomes are structured and how they have evolved. Clear mapping data as well as some preliminary sequence data show the extent of synteny, i.e. conservation of gene order, between genomes of plants from the same family or from related ones. This has been especially documented for monocotyledonous species in the grass family. Through identification and mapping of synteny, it will be possible to isolate genes from crop plants with large genomes, using information on the homologous genes in a related plant with a smaller genome, such as Arabidopsis. Because of the importance of some of the domesticated grasses, such as rice, wheat, and maize, to the human food chain, developments in this area can lead directly to opportunities to improve the productivity of our food systems.

Functional Genomics

Monitoring Gene Expression

Information on where and when a certain gene is expressed can provide indications of the biological function of the encoded protein. Classical approaches such as Northern blot analysis allow such a study on a single gene, but in view of the huge amount of data from the genome programs, from plants as well as from other organisms, strategies have been developed to look at expression on a genome-based level.

DNA Micro-arrays

In principle DNA micro-arrays are a kind of 'reverse Northern blot' whereby DNA clones (cDNA clones, PCR-generated fragments, etc.) are spotted in a dense array and hybridized to RNA-derived probes. The hybridization signal can be quantified automatically and reflects the abundance of the corresponding mRNA in the total RNA pool. The value of this technique is in the miniaturization, whereby large numbers of clones can be analyzed simultaneously. Three techniques are currently in use: (1) DNA spotted on nitrocellulose filters and hybridized to radioactive probes, (2) DNA spotted on coated glass slides and hybridized to fluorescently labeled probes, and (3) DNA oligonucleotides synthesized in situ on a solid support and hybridized to fluorescent probes. This last technique is mostly referred to as DNA chips and has been developed by Affimetrix (Santa Clara, CA, USA).

To be consulted and queried by the scientific community, hybridization data of the micro-arrays should be stored in public expression databases and linked to sequence and mutant database entries with gene identity and structure. Particularly useful information, such as insights into gene regulation and interaction networks, will come from the clustering of the genes with the same expression profile and from the expression study of mutants.

Differential Display

This method makes it possible to analyze and compare transcribed genes systematically in a bi-directional fashion in a one-tube reaction. It involves the isolation of RNA from the tissues to be compared, followed by PCR amplification of this RNA using random primers. Samples are then analyzed by gel electrophoreses and the patterns of the amplified cDNAs compared. cDNAs that are found to be differentially amplified can then be eluted from the gel and cloned. The most significant advantages of differential display are its simplicity, the small amount of RNA needed as start material, and the possibility of detecting virtually all differentially expressed mRNAs if a large number of primer combinations are performed. Disadvantages, however, are the problems with reproducibility and the reliability of the observed differences, because Northern blot data of the identified genes do not always confirm the data obtained by differential display.

Transcript Profiling

Transcript profiling is also referred to as cDNA-AFLP. It is a technique based on AFLP, modified for use on mRNA material. The principle is comparable to that of differential display, but the cDNA material is first cut with restriction enzymes, after which an adapter is annealed to these sites. Primers based on this adapter sequence can then amplify the cDNA, making use of extra selective nucleotides. When transcript profiling is done with all the possible primer combinations it should, in theory, yield information on about 80% of the transcripts, depending on the enzymes that are used in the protocol. In addition, if the differentially expressed AFLP bands are sequenced, one would be able to link expression data with the available genome information.

Gene Knock-out and Mutagenesis

The ultimate tool for studying gene function is gene knock-out via homologous recombination. Unfortunately in plants very little success has been realized with target-specific constructs for gene replacement. Only for a moss, Physcomitrella patens, good results have been obtained.

To overcome the lack of a good knock-out system, a random mutational approach has been widely used and seems to be very promising in the field of plant functional genomics. Two classes of mutants can be distinguished: firstly, mutants generated by mutagens, such as ethyl methanesulfonate and X-rays and, secondly, insertional mutant collections whereby a piece of foreign DNA is randomly inserted into the genome.

Classical Mutagenesis and Map-based Cloning

This approach has long been the search for a needle in the haystack. Indeed, once a certain mutant with an interesting phenotype was obtained by random mutagenesis it was difficult and at least time-consuming to map the position of the gene and 'walk' or 'land' on it. However, recent techniques, such as AFLP, bulk segregant analysis, and the available maps and sequencing data allow a fast mapping of mutants.

Insertional Mutagenesis

As the identification of classical mutants has long been a time-consuming effort, an approach has been followed to introduce foreign pieces of DNA into the plant genome to interrupt genes at random. The T-DNA of Agrobacterium as well as transposons have been used. Because the sequence of the inserted element is known, these libraries can be screened by PCR-based strategies built on sequence information of the gene in which one wants to find a mutant. For instance, theoretically for every gene of Arabidopsis that has been identified within the sequencing program, an insertional mutant should be available in one of the world-wide collections.

Bioinformatics

The common feature in any genome-wide approach is the production of enormous amounts of raw data. Bioinformatics was soon seen as a way to help produce and deal with these data in specific databases. Nevertheless, the need to transform these data into information that biologists can use and query properly has probably been underestimated. As an effort toward building such knowledge, computer tools have been developed to decipher the gene architecture of the Arabidopsis genome and provide annotation of the genomic sequences. However, distinguishing coding from non-coding sequences and setting the proper gene structure remains a problem. The necessary EST/cDNA information is far from complete and the performance of computer prediction tools that are used instead has not been evaluated yet. More importantly, a proper computer-assisted environment for whole-genome annotation is still missing. This lack largely explains why the present-day annotation is so faulty, limiting its use for data mining and experiment setting. The progress in functional and expression studies will increase the challenge to handle all data. It will certainly help solving the functional annotation problem, if the proper effort is made to settle an adequate ontology and develop the tools to analyze the data and turn them into knowledge on gene biological function that is properly linked to gene structure.

SEQUENCING AND ASSEMBLING A PLANT GENOME

Sequencing and assembling a complete plant genome has been seen as a daunting task. Indeed, the first plant genome of Arabidopsis thaliana. The current generation of DNA sequencing technologies, however, is making genome sequencing a reality even for small labs without generous funding sources. This momentous leap brings exciting opportunities to the botanical community. Whole genomes, paired with resequencing, can provide thousands of nuclear markers for phylogenetic and population-level studies, enabling genome-wide investigations into fundamental evolutionary and ecological questions. In addition, generating a pan-genome capturing the genomic diversity of ecotypes, geographical isolates, and related species will make comparative approaches and association studies possible to identify the genetic components of certain traits and adaptations. The possibilities run the gamut from systematics, ecology and evolution, to molecular genetics.

Despite the dramatic drop in sequencing cost and the rise in throughput and read length, care still needs to be taken when planning a genome project in order to maximize assembly quality versus cost.

Not all plants are equally sequenceable. Genome size, repeat structure and age, and heterozygosity are the three main factors that determine the feasibility of the project. In order to strategize the sequencing approach, certain groundwork is necessary.

Genome Size and Complexity

Plant genome sizes vary dramatically, ranging from 0.063 to 148.8 Gbp, and the sequencing cost increases as the genome size increases. Indeed, only a few genomes larger than 10 Gbp have been assembled, such as wheat, Ginkgo L., Picea A. Dietr., and Pinus L. In addition, assembly of allo- or autopolyploid genomes is complicated by the presence of additional haplotypes. Therefore, identifying a haploid or diploid individual with a relatively small genome in your clade of interest is critical; this can not only save significant amounts of money, but also simplify downstream bioinformatics analyses. However, if such individuals are not available or if polyploids are actually the targets, one should consider long-read sequencing coupled with Hi-C, optical mapping, or 10× Genomics.

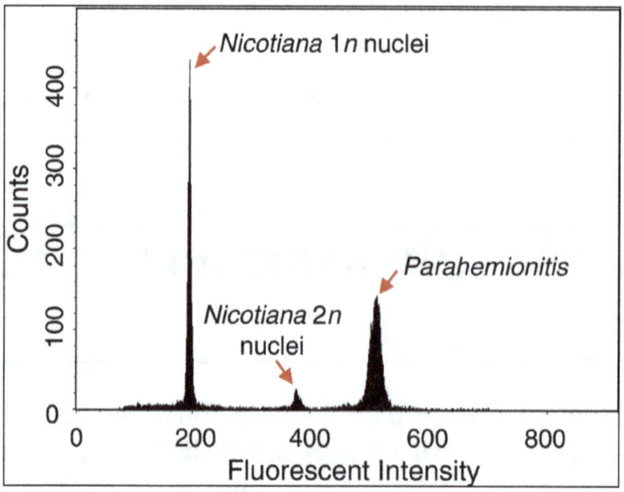

An example flow cytometry result, using Nicotiana as the standard to infer the genome size of the fern Parahemionitis cordata.

Flow cytometry is a common and accurate way to determine genome size, but it requires fresh material and buffer optimization. For lineages rife with polyploidy, it is important to determine the ploidy level, either based on chromosome squash or by measuring pollen, stomata, or spore size.

In addition to genome size, heterozygosity is another important consideration. When assembling short-read shotgun sequences, heterozygous regions complicate graph structure and make it difficult to phase haplotypes. One way to reduce heterozygosity is to create inbred lines or doubled haploids, but this is time-consuming and, furthermore, not every plant can be selfed for several generations or passed through anther culture and regenerated. For gymnosperms, inbreeding for just one generation might take decades, but certain species have large mega-gametophytes that might provide enough haploid DNA for sequencing. Some ferns, on the other hand, are capable of intra-gametophytic selfing, which creates a complete homozygous sporophyte in one generation and can be effectively treated as a haploid in genome sequencing.

K-mer Frequency Distribution

To discern if the chosen individual is appropriate for genome sequencing, a powerful and simple approach is to use raw Illumina DNA shotgun reads (Illumina, San Diego, California, USA) to infer genome size, repeat percentage, and heterozygosity. A k-mer distribution refers to all the possible subsequences of length k that are contained within a string (or a set of strings) of nucleotides, such as a genome assembly or collection of sequencing reads. Figure 2A shows a hypothetical k-mer frequency plot (K = 31) from shotgun Illumina sequencing data. The x-axis shows the number of times a given k-mer (e.g., ATGCTAGCTAACTAGACTACTAAGCTAGCAT) appears in the Illumina reads, and the y-axis shows the number of unique k-mers at that frequency. For example, the red arrow marks where 5 million unique k-mers are found exactly 20 times. The first peak close to 1 (blue arrow) is largely due to sequencing errors that created abundant unique k-mers of low frequency. The second peak (red arrow) represents the sequencing coverage, meaning that the majority of k-mers were sequenced 20 times. The genome size can then be approximated by the total number of k-mers (the area under the curve), divided by sequencing coverage.

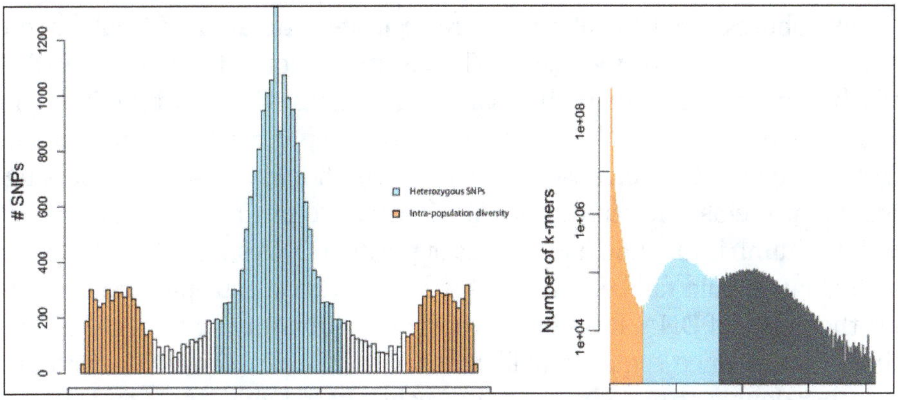

Hypothetical k-mer frequency plots, showing a low-heterozygosity genome (A) and a highly heterozygous and repetitive genome (B).

A k-mer frequency plot can also be used to estimate the level of heterozygosity in an individual. K-mers from a heterozygous site will have half of the sequencing coverage compared to the homozygous region, creating an intermediate peak or shoulder half-way to the genome coverage peak. The higher this peak (or shoulder), the higher the sample heterozygosity. Similarly, k-mers from repetitive regions will have a much higher representation than the mean coverage, and show up as a high frequency shoulder toward the right of the frequency plot. K-mer plots can be generated by the Jellyfish package or KmerGenie, from where a few lines of R script can then be used to estimate the genome size. It should be noted that the sequencing depth has to be sufficiently high, at least 30× coverage or more, in order for peaks to emerge.

The percentage and content of repetitive elements in a genome can also be inferred by shotgun Illumina DNA sequencing data before assembly. Sequence reads that cover as little as 1% of the genome size can be used as input to programs like RepeatExplorer or Transposome. Random, low-coverage shotgun reads are clustered by nucleotide similarity and overlap, then annotated against a plant repeat database like Repbase or one derived from a closely related genome.

Given the relatively low cost of Illumina sequencing and its usefulness in several ways before and after genome assembly, generating Illumina shotgun DNA sequence data is a critical step when starting a genome assembly project.

DNA Quality and Quantity

The quantity, purity, and integrity of DNA often dictate the quality of the final genome assembly and therefore should not be ignored. For long-read sequencing with Pacific Biosciences (or "PacBio") sequencing systems (Pacific Biosciences, Menlo Park, California, USA), over 10 μg of DNA with 30–50-kbp average fragment size is usually needed to obtain optimal results with the current chemistry. To obtain the best DNA possible, one could use the classic nuclei preparation protocols, although grams of starting plant materials are necessary. QIAGEN MagAttract and Bionano IrysPrep) are kit-based alternatives. The integrity of DNA can be roughly and quickly visualized by conventional agarose electrophoresis, but to obtain a more accurate estimation of DNA fragment size, pulse-field electrophoresis or the Agilent TapeStation should be used. The DNA purity should also be high, as contamination of salts and proteins will likely inhibit enzyme activity during library preparation. The NanoDrop or other spectrophotometers that can measure 260/280 and 260/230 nm absorbance readings should be used for assessing sample purity, as DNA, proteins, and salts absorb at 260 nm, 280 nm, and 230 nm, respectively. A general rule of thumb is that 260/280 values of purified DNA should fall between 1.8–2.0, whereas 260/230 should fall between 2.0–2.2. Although spectrophotometry is useful to determine the purity of DNA, it is not appropriate for quantifying total genomic DNA for sequencing or for purified sequencing libraries. Qubit Broad Range (Thermo Fisher Scientific) or other double-stranded DNA fluorescence-based quantification is the preferred DNA quantification method for total genomic DNA.

Sequencing Platforms

There are two major types of sequencers, one that generates short reads (<300 bp) massively, accurately, and cheaply (e.g., Illumina and BGISEQ-500), and a second type that produces longer reads (>10 kbp) but inaccurately (7–12% error rate) and with a much lower throughput (e.g., PacBio and Oxford Nanopore). For de novo sequencing, the PacBio or Nanopore long-read technology is preferred and can be realistically done for a small- to moderate-sized genome (<1 Gbp). Generating sufficient long-read coverage for larger genomes might be prohibitively expensive, and a hybrid approach combining both short and long reads would be more feasible.

Illumina

The Illumina platforms have been the main workhorse for genome sequencing; they are able to create massive sequencing data cheaply and with a low error rate. Currently one lane of Illumina HiSeq4000 costs approximately US$2500, and outputs 90 to 100 Gbp of paired-end 150-nucleotide reads. For a genome of 1 Gbp, this throughput translates into impressive 100× coverage. The downside, however, is that the read length is short: 150 nucleotide maximum for HiSeq and 300 nucleotide for MiSeq. In addition, the library insert size usually cannot go beyond 800 nucleotides, providing only short-range information. Therefore, for de novo sequencing, additional scaffolding approach is needed to achieve a reasonably good assembly.

BGISEQ

BGI, a prominent sequencing provider and biotech company in Shenzhen, China, recently unveiled its own sequencer, BGISEQ-500. This sequencer is based on Complete Genomics' nanoball technology for creating sequencing clusters, which is different from the bridge amplification method used in Illumina. The specification of BGISEQ-500 is reportedly comparable to Illumina HiSeq2500, with similar read length, throughput, and error rate. This platform has not yet been widely adopted, however, and the pros and cons of BGISEQ-500 over Illumina are still to be determined.

PacBio

Single-molecule sequencing such as PacBio and Oxford Nanopore can read long DNA molecules (>10 kbp) without prior amplification. It is thus extremely useful for de novo sequencing of plant genomes. The disadvantage, however, is that the current single-molecule sequencing technologies all have a high error rate ranging from 10–15% and come with a higher price tag per base than Illumina. One PacBio Sequel SMRT cell (2.0 chemistry and 10-h movie) costs roughly US$1250 and yields 5 Gbp of long-read data (>10–20 kbp). Given that at least 40× coverage is recommended for PacBio-only assembly, a 1-Gbp genome will cost US$10,000. It should be stressed that DNA quality, particularly the fragment length, matters tremendously for PacBio (and also for Nanopore) sequencing.

Oxford Nanopore MinION

MinION is similar to PacBio in terms of read-length, throughput, and error rate, but the entire sequencer is packaged into a USB-sized device that is highly portable. The portability stands out among all the sequencers all it requires is a moderately modern laptop and the library preparation can be completed in as little as 10 min. Because of this, MinION has been used in the International Space Station, as well as in arctic and other remote research stations. The genome of Solanum pennellii Correll (1–1.1 Gbp) was recently assembled based entirely on Nanopore data, and Michael et al. reported that an Arabidopsis thaliana genome (135 Mbp) could be de novo sequenced by just one MinION flow cell. It is therefore possible to DNA barcode, genotype, or even sequence entire plant genomes real-time in the field. The era of "mobile genomics" might soon be coming, although the hurdle now is how to efficiently extract high-quality DNA outside of laboratories.

Oxford Nanopore is rapidly evolving both in terms of scalability and library preparation methods. For instance, the available GridION system is able to concurrently run up to five of the MinION-sized flow cells with integrated computing power. Flow cell pricing is currently as low as US$300 each when purchased alongside the capital cost of the machine. Each of these five GridION flow cells are regularly able to generate more than 5 Gbp of long-read data, with throughput varying with different DNA input quality, size selection, and library types. Library preparations for DNA vary from 5-min transposase-based rapid kits (longer reads, but lower throughput) to more traditional ligation-based preparations (shorter reads, but maximum throughput). Reads can also be generated from a single strand of DNA (1D sequencing) or from a newly released library preparation that consecutively sequences both strands of a complementary DNA molecule ($1D^2$ sequencing) for improved accuracy.

Genome Scaffolding Approaches

To improve genome assembly, it is critical to obtain long-range information to orient and order contigs into scaffolds. Traditionally, this has been done either by generating a physical map or by constructing and sequencing bacterial artificial chromosome (BAC) libraries, both of which are laborious and costly. Fortunately, there are a few clever library preparation methods and new technological advances that make genome scaffolding much more cost-effective and feasible for small labs.

Mate-pair Library

This library preparation method takes long DNA fragments (>1 kbp) and self-ligates them into circles, bringing the distant ends together. The circular DNAs are then cut, and the short fragments containing the joined junction are selected. From these fragments, the two distant parts of the genome can then be sequenced by standard Illumina paired-end sequencing. Mate-pair libraries, however, are not easy to construct, requiring a large quantity of high-molecular-weight DNA, and usually have low complexity (i.e., with many duplicates), thereby wasting the sequencing output. Given that PacBio and Nanopore read lengths regularly exceed 10–30 kbp in length, Illumina mate-pair libraries are quickly falling out of favor.

Hi-C

In eukaryotic cells, nuclear DNA wraps around histones and is packed into a complex three-dimensional chromatin conformation. Within this structure, two DNA strings packed in close proximity might be coming from two distant regions of chromatin, and such spatial relationships can be leveraged to create long-range sequence information. The Hi-C method first cross-links DNA to histones in vivo to preserve the chromatin conformation. DNA is then digested, allowing spatially close, but physically distant, DNA fragments to be ligated. The resulting libraries can be sequenced on the Illumina platform. Hi-C has been shown to be able to create chromosomal-level genome assemblies given a draft assembly and the expected chromosome number, the Hi-C scaffolder uses a statistical model to piece contigs together into individual chromosomes. In addition, native plant chromosomes are unlikely to be cross-linked with foreign DNA, and contaminations can therefore be identified and excluded. The most notable feature of Hi-C is that it does not require high-molecular-weight DNA (the cross-linking, digestion, and ligation are done in vivo), thus providing a solution to materials that resist DNA extraction. Dovetail Genomics and Phase Genomics offer commercial solutions to Hi-C sequencing and scaffolding, often at a total cost between US$10,000 and US$20,000. Dovetail Genomics also has a Hi-C variant, called the Chicago library; instead of cross-linking in vivo, this method applies exogenous histones to pure DNA extraction to recreate chromatin structure in vitro but produces shorter inserts than in vivo Hi-C. Recently, in vitro and in vivo Hi-C have been used to improve numerous plant genomes, including lettuce, barley, amaranth, and quinoa.

BioNano Optical Mapping

The goal of optical mapping is to create a genomic restriction map, thereby providing a backbone to scaffold the contigs. It starts with ultra-high-molecular-weight DNA (>150 kbp) and uses single-stranded restriction endonucleases to create nicks at specific recognition sites. Fluorescent nucleotides are then incorporated at each of the nicked sites. To visualize the labeling pattern, the DNA is applied onto a chip, where each molecule enters a nanochannel and becomes linearized. The intervals between fluorescent labels can be imaged precisely and in a high-throughput fashion, thus enabling the construction of a genome-wide restriction map. This restriction map can be paired with an in silico digestion of the genome assembly, allowing for scaffolding and assembly correction. Notably, multiple enzymes can be used to produce separate but complementary optical maps. The “two enzyme” approach mitigates a previous limitation of the technology, where nearby restriction sites on opposite strands would lead to DNA molecule breakage. Several high-quality plant genomes have recently been scaffolded and improved with optical mapping, including maize, garden asparagus, and Oropetium Trin, and the technique is even feasible for individually flow-sorted chromosomes.

10× Genomics

At the heart of 10× Genomics is their Chromium microfluidic controller. This platform, which resembles a toaster, enables massive partitioning of the input genomic DNA into oil droplets (a dozen or so DNA molecules per droplet), within which library preparation is performed and, importantly, each partition receives a unique barcode. Because reads with the same barcode can only come from one of the few DNA molecules, the long-range information can be bioinformatically deduced to produce "linked" reads. In addition, homologous (or homeologous) DNA molecules are unlikely to be included in the same droplet, so that each haplotype will receive its own barcode and can be distinguished. For a highly heterozygous or polyploid individual, 10× Genomics linked reads can potentially provide a phased genome. Like many other scaffolding approaches, the length of DNA molecules determines the quality of long-range information. The advantages of 10× Genomics technology, however, are that it only requires 1 ng of input DNA and is a more economical way to obtain long-range information than Hi-C or optical mapping.

Affordable personal genomes have been a motivation for the development of low cost, high-throughput next-generation sequencing (NGS) technologies, including Roche/454, Illumina and AB SOLiD, which are able to generate three to four orders of magnitude more DNA sequence than Sanger-based sequencing [a first-generation sequencing (FGS) technology] on the ABI 3730xl platform. These NGS technologies have enabled the genomics community to comprehensively characterize DNA sequence variation within a species by sequencing multiple accessions/genotypes, de novo sequencing of a number of species, detection of methlylated regions in genome and gene expression profiling.

In the past, Sanger sequencing has been used to characterize the genomes of several organisms including model plants as well as major crop species like rice, soybean, sorghum, maize, grape and eucalyptus. The availability of NGS technologies, however, have enabled the research community to embark upon sequence the genomes of thousands of plant species through the undertaking of the 1000 Plant Genomes Project the 1001 Arabidopsis Genome project and the 1000 Plant and Animal Genome Project. Similarly, the Genome 10 K Project has been conceived to sequence and assemble 10 000 vertebrate genomes including at least one from each genus.

Advances in nanobiology and robotics for DNA sequencing applications have also been driven by a competition to win the race of sequencing a human genome at a target price of US $1000 and therefore new sequencing technologies and platforms continue to emerge. As a result, existing NGS technologies are referred as second-generation sequencing (SGS) technologies and future or very recently available NGS technologies are referred to as third-generation sequencing (TGS) or 'next-next' generation sequencing (NNGS) technologies. This article provides an overview of different sequencing technologies with a major emphasis on the forthcoming TGS technologies. We also discuss different bioinformatics tools required for data analysis of the massive amounts of sequence data emerging from these technologies.

FGS Platforms

The FGS methods include Sanger's enzymatic dideoxy DNA sequencing and the Maxam and Gilbert's chemical degradation methods. For commercial DNA sequencing, Applied Biosystems was the first company to introduce ABI Prism 377 based on slab gel electrophoresis. Owing to the inconvenience of casting gels, the ABI Prism 3700 was developed with automated reloading of the 96 capillaries with a polymer matrix. This platform was used in the sequencing of the first human genome. Sanger sequencing was also used for sequencing genomes of several plant species such as Arabidopsis (Arabidopsis thaliana), rice (Oryza sativa), sorghum (Sorghum bicolor), grapes (Vitis vinifera), poplar (Populus trichophora) and soybean (Glycine max). Sequencing time and personnel costs associated with Sanger sequencing, however, prohibited the sequencing of a large number of plant species, especially those with large, complex genomes (e.g. wheat, 16 Gb).

SGS Platforms

In 2005, 454 Life Sciences launched the GS 20, the first NGS systems into the market. After acquiring 454 Life Sciences, Roche Applied Science extended this technology to the new version of the 454 instrument, the GS FLX titanium. Subsequently, Roche/454 launched several other platforms including GS 20/FLX, GS FLX Titanium+, GS FLX Titanium XLR70 and GS Junior. In parallel, several other companies launched competing NGS systems that included 'Solexa 1G' (later named 'Genome Analyzer'), GA, GA II, HiSeq 2000, HiSeq 1000, Hi ScanSQ and MiSeq by Illiumina Inc. SOLiD 3 and SOLiD 4 system by Applied Biosystems Recently a new system for NGS based on multiplex polony technology named as the Polonator G.007 has been introduced by Dover and Harvard Medical School Currently, these technologies are referred as SGS systems. Although all these systems can be used for a multitude of applications for plant genomics research, Illumina and Roche/454 have been the most widely adopted SGS platforms as evident by publications.

Illumina and Roche/454 employ the principle of sequencing by synthesis (SBS) i.e. they rely on PCR to amplify a given DNA template which is then attached to a solid surface and are subsequently imaged in a phased approach. On the other hand, sequencing platforms like SOLiD 3 and SOLiD 4 employ sequencing by ligation (SBL). However, the amplification process can introduce errors in the template sequence as well as introduce amplification bias. In addition, generation of NGS data takes several days due to a large number of instrument scanning and washing cycles. Because of dephasing, as compared to Sanger sequencing, average read length of sequence reads produced by SGS platforms is shorter. Based on the throughput achieved, except MiSeq, all other sequencing platforms from Illumina (GA, GA II, HiSeq 2000, HiSeq 1000 and Hi ScanSQ), Applied Biosystems (SOLiD 3 and SOLiD 4) and Roche/454 (GS 20/FLX, GS FLX Titanium+, GS FLX Titanium XLR70 and GS Junior) are considered as high-throughput SGS platforms.

The GS FLX from 454 Life Sciences produces over a million reads of up to 1000 bases per 10 h run, for a total yield of 400–600 megabases. Thus, 454 Sequencer has longest short reads among all SGS platforms. The Illumina Genome Analyzer yields over one hundred million high-quality short reads (up to 76 bases) per 3–5 day run, totaling several gigabases of aligned sequence. At present, the new Illumina HiSeq 2000 Genome Analyzer is capable of producing single reads of 2 × 100 bp (pair-end reads), and generates ~200 Gb of short sequences per run. The raw base accuracy is >99.5%. Finally, the Applied Biosystems SOLiD system also produces hundreds of millions of short reads (up to 50 bases) per run.

The large amount of data generated by these high-throughput SGS technologies poses a challenge for data storage and transfer and informatics operations. This is especially true for the shorter reads generated by the Illumina and SOLiD systems that make sequence alignment and assembly processes challenging. Nevertheless, SGS technologies are being used for a variety of applications including de novo sequencing of genomes, transcriptome analysis, gene expression, marker discovery and many others in plant species such as cocoa, chickpea, pigeonpea, date palm and pea.

TGS Technologies

In the context of challenges associated with assembling of short sequence reads, development of technologies that generate longer sequence reads will help to deliver the information required for assembling complex genomes. In addition, as SGS platforms generally require either an in vitro or in vivo amplification step, technologies those that directly sequence single molecules of DNA, eliminating the need for costly and many times problematic procedures like cloning and PCR amplification are preferred. To this end, a number of academic and commercial efforts are developing ultra-low-cost 'single-molecule' sequencing (SMS) technologies. SMS technologies can be grouped into three categories:

- fluorescence-based methods for SMS like exonucleolytic degradation, true single-molecule sequencing (tSMS), fluorescence resonance energy transfer (FRET)-based approach, single-molecule real-time sequencing (SMRT) and microfluidic devices.
- non-fluorescent sequencing systems like Nanopore's Nano-edges, sequencing using transmission electron microscopy, pyrosequencing, motion-based sequencing and scanning tunneling spectroscopy-based sequencing.

- Raman-based methods such as sequencing using surface-enhanced Raman spectroscopy (SERS) and sequencing using tip-enhanced Raman spectroscopy (TERS), these technologies are referred as TGS. TGS technologies seem to be superior to SGS technologies as they generate longer sequence reads in higher throughput fashion and faster turnaround time with higher consensus accuracy.

Some of these platforms that have potential for extensive use in plant genomics research are given below.

SMRT Sequencer

Pacific Biosciences company has recently introduced the PacBioRS, the TGS system that employs the SMRT DNA sequencing technology, where in DNA sequencing is performed on SMRT cells (nanofabricated consumable substrates). In this technology, DNA fragment is sequenced by a single DNA polymerase molecule that is attached to the bottom of each zero-mode waveguides (ZMW) and as a result, each DNA polymerase resides at detection zone of ZMW. As per the company, PacBioRS requires less than a day from sample preparation to obtaining the sequence information and produces read lengths > 1000 bp. The SMRT sequencer has been available to several sequencing centers and critical assessment on its performance is ongoing.

Heliscope Single Molecule Sequencer

This sequencer has been introduced by Helicos company that images billions of single molecules and produces 21–35 Gb per run, almost 100X greater than Sanger methods, and faster than many currently available NGS technologies. HeliScope employs true single-molecule sequencing (tSMS) chemistry and direct RNA sequencing chemistries. Large numbers of strands of single DNA molecule can be sequenced simultaneously by using tSMS chemistry. tSMS has been used to sequence an individual human genome, re-sequence the M13 virus genome and to quantify the yeast transcriptome. A drawback is a relatively high raw sequence error rate that can be overcome with repetitive sequencing, but increases the cost per base for a given accuracy rate, offsetting some of the gains from lower reagent costs.

Ion Personal Genome Machine Sequencer

Life Technologies company has recently launched Ion Personal Genome Machine (PGM) Sequencer based on the ion torrent semiconductor technology. This technology is based on a biochemical process by which a hydrogen ion is released as a nucleotide and is incorporated into a strand of DNA by a polymerase. This technology is independent of enzymatic reactions, fluorescence, chemi-luminiscence, and optics. It uses a high-density array of micro-machined wells. Each of these wells holds a different DNA template. Just beneath the wells, there is an ion-sensitive layer and a proprietary Ion sensor. During the sequencing, when a new base is added to the template, and incorporated into the strand, hydrogen ion is released. The charge from that ion changes the pH of the solution that can be detected directly by the ion sensor without imaging. The PGM system can perform a wide range of sequencing applications including multiplexing amplicons, transcriptome analysis, small RNA discovery, and ChIP-Seq analysis.

Bioinformatics Tools

The increase in sequence throughput from different sequencing platforms is exponential. In this context, storage and management of humongous datasets is very challenging. In addition to data storage and management, primary, secondary and tertiary analysis solutions like quality control, base calling, de novo assembly, alignment to a reference genome, variant calling, Chip-Seq, transcriptome analysis are necessary to make sense of the larger volumes of sequence data. As existing sequence analysis tools were not appropriate for analysis of sequence data coming out from new sequencing technologies, a number of tools/software packages have been developed in last few years. The bioinformatics community needs to be ready continually develop new tools as well as data storage and management systems in anticipation of even larger amount of sequence data coming out from TGS technologies. Moreover, the types of data and quality associated with each will complicate analyses and the use of existing tools. Cloud computing is a potential solution to the question of massive data storage as well as analysis. Cloud computing provides computation, software, data access and storage services that do not require end-user knowledge of the physical location and configuration of the system that delivers the services.

PLANT GENOME ASSEMBLY

A plant genome assembly represents the complete genomic sequence of a plant species, which is assembled into chromosomes and other organelles by using DNA (deoxyribonucleic acid) fragments that are obtained from different types of sequencing technology.

Structure

The genome of plants can vary in their structure and complexity from small genomes like green algae (15 Mbp) to very large and complex genomes that have typically much higher ploidy, higher rates of heterozygosity and repetitive elements than species from other kingdoms. One of the most complex plant genome assemblies available is that of loblolly pine (22 Gbp). Due to their complexity, the plants' genome sequences can't be assembled back into chromosomes using only short reads provided by next-generation-sequencing technologies (NGS), and therefore most plant genome assemblies available that used NGS alone are highly fragmented, contain large numbers of contigs, and genome regions are not finished. Highly repetitive sequences, often larger than 10kbp, are the main challenge in plants. Most of the chromosomal sequences are produced by the activity of mobile genetic elements (MGEs) in the plant genomes. MGEs are divided into two classes: class I or retrotransposons, and class II or DNA transposons. In plants, long- terminal repeat (LTR) retrotransposons are predominant and constitute from 15% to 90% of the genome. Polyploidy is another challenge in assembling a plant genome, and it is estimated that 80% of plants are polyploids.

Assemblies

The first complete plant genome assembly, that of Arabidopsis thaliana, was finished in 2000, being the third multicellular eukaryotic genome published after C. elegans and D. melanogaster. Arabidopsis, unlike other plants' genomes (e.g. Malus) has convenient traits, such as a small nuclear genome (135Mbp) and a short generation time (8 weeks from seed to seed). The genome has five chromosomes reflecting approximately 4% of the human genome size. The genome was sequenced and annotated by the Arabidopsis Genome Initiative (AGI).

The initiative for sequencing the genome of rice (Oryza sativa), began in September 1997, when scientists from many nations agreed to an international collaboration to sequence the rice genome, forming "The International Rice Genome Sequencing Project" (IRGSP). At an estimated size between 400-430 Mb, approximatively four times larger in dimensions than A. thaliana, rice has the smallest of the major cereal crop genomes.

Between 2000 and 2008 in total 10 plant genomes were published while in 2012 alone, 13 plant genomes were published. Since then the number was constantly increasing, and now more than 400 plant genomes are available in the NCBI genome database, of which 72 were re-annotated (NCBI).

Databases

EnsemblPlants is part of EnsemblGenome database and contains resources for a reduced number of sequenced plant species. It mainly provides genome sequences, gene models, functional annotations and polymorphic loci. For some of the plant species, additional information is provided including population structure, individual genotypes, linkage, and phenotype data.

Gramene is an online web database resource for plant comparative genomics and pathway analysis based on Ensembl technology.

Plant Genome DataBase Japan PGDBj) is a website that contains information related to genomes of model and crop plants from databases. It has three main components: ortholog db, DNA marker and linkage map db, and plant resource db, where multiple plant resources accumulated by different institutes are integrated. The aim is "to provide a platform, enabling comparative searches of different resources".

PlantsDB is a resource for analysing and storing genetic and genomic information from various plants, and offers tools to query these data and to perform comparative analysis with the help of in-house tools.

PLAZA is another online resource for comparative genomics that integrates plant sequence data and comparative genomic methods, and performs evolutionary analysis within the green plant lineage (Viridiplantae).

Assembly Strategies

In general, for sequencing and assembling large and complex genomes like plants, different strategies are used, based on the technologies available at that time when the project started.

Sanger Clone-by-clone

Clone-by-clone sequencing strategies are based on the construction of a map for each chromosome before the sequencing, and rely on libraries made from large-insert clones. The most common type of large-insert clone is the bacterial artificial chromosome (BAC).

With BAC, the genome is first split into smaller pieces with the location recorded. The pieces of DNA are then inserted into BAC clones that are further multiplied by inserting them into bacterial cells that grow very fast. These pieces are further fragmented into overlapping smaller pieces that are placed into a vector and then sequenced. The small pieces are then assembled into contigs by overlapping them. Next, using the map from the first step the contigs are assembled back into the chromosomes.

The first complete plant genome assembly (also the first plant genome published) that used this type of technique was Arabidopsis thaliana, in 2000. Different large-insert libraries like BACs, P1 artificial chromosomes (PAC), yeast artificial chromosome (YAC) and transformation-competent artificial chromosomes (TACs) were combined to assemble the genome. From clones with restriction fragment fingerprint, by comparison of the patterns and hybridization or polymerase chain reaction (PCR) the physical maps were constructed. The physical maps were integrated together with genetic maps to identify contig positions and orientations. End sequences from 47,788 BAC clones were used to extend contigs from anchored BACs and to select a minimum tiling path. A total of 1,569 clones found in minimum tiling path were selected and sequenced. Direct PCR products were used to clone remaining gaps, and YACs allowed the characterization of telomere sequences. The resulting sequenced regions were 115.4 Mb of the 125 Mb predicted size of the genome and a total of 25,498 of protein-coding genes.

To sequence and assemble the genome of Oryza Sativa (japonica), the same strategy was used. For Oryza Sativa a total of 3,401 mapped clones in a minimum tiling path were selected from the physical map and assembled.

One of the most important crops in the world, maize (Zea mays), is the last plant genome project primarily based on Sanger BAC-by-BAC strategy. The genome size of Maize, 2.3 Gb and 10 chromosomes, is significantly larger than that of rice and Arabidopsis. To assemble the genome of maize a set of 16,848 minimally overlapping BAC clones derived from combinations of physical and genetic map were selected and sequenced. The assembly on maize was performed in addition with external information data. The data was obtained from cDNA and sequences from libraries with methyl-filtered DNA (libraries that uses the knowledge that the bases in genic sequences tends to be less heavily methylated than those in non-genic regions) and high Co t techniques.

Sanger clone-by-clone strategy has the advantage of working in small units, which reduces the complexity and computational requirements, as well as minimized problems associated with the misassembly of highly repetitive DNA and therefore is an attractive solution in assembling plant genomes and other complex eukaryotic genomes. The main disadvantages of this method are the costs and the resources required. The cost of the first plant genome assemblies was estimated between 70 million dollars and 200 million dollars per assembly.

Sanger Whole-genome Shotgun

In the WGS sequencing technology there is no order for the fragments that are sequenced. The DNA is randomly sheared and cloned fragments are sequenced and assembled using computational methods. This technology reduced the cost and the time associated with construction of the maps and relies on computational resources.

A considerable number of important plant genomes like grapevine (Vitis Vinifer), papaya (Carica papaya), and cottonwood (Populus trichocarpa) were sequenced and assembled with Sanger WGS strategy.

The draft genome of grapevine is the fourth genome published for a flowering plant and the first from a fruit crop. The sequences of the genome were obtained from different types of libraries, like plasmids, fosmids and BACs. All the data were generated by paired-end sequencing of cloned insert using Sanger technology on ABI3730x1 sequencers. To assemble the reads, Arachne, a software designed to analyze reads obtained from both ends of plasmid clones, was used. In total 6.2 million paired-end tag reads were produced. The software produced 20.784 contigs that were combined into 3,830 supercontigs, having an N50 value of 64kb. Supercontigs had a total size of 498 Mb.

The anchorage of the supercontigs along the genome was performed first by joining supercontigs together using paired BAC end sequences. The resulting ultracontigs and the remained supercontigs were then aligned along the genetic map of the genome. Later improvements of this strategy enabled the sequencing of Brachypodium distachyon, Sorghum bicolor and soybean.

Next-generation Sequencing

Due to its relatively cheap cost in comparison to previous methods, most of the recent plant genomes were sequenced and assembled using data from NGS (next-generation-sequencing) technology. In general the NGS data are used in combination with Sanger Sequencing technology or long-reads obtained from the third generation sequencing. The genome of the cucumber, (Cucumis sativus), was one of the plant genomes that used the NGS Illumina reads in combination with Sanger sequences. 72.2-fold genome coverage high quality base pairs were generated from which 3.9-fold coverage was provided from Sanger and the Illumina GA reads provided 68.3-fold coverage. From these two assemblies were produced based on the sequencing technology. The resulting contigs were compared between them, resulting in a total length of the assembled genome of 243.5 Mb. The result is about 30% smaller than the genome size estimated by flow cytometry of isolated nuclei stained with propidium iodide (367 Mb). A genetic map was constructed to anchor the assembled genome. 72.8% of the assembled sequences were successfully anchored onto the seven chromosomes. Another plant genome that combined NGS with Sanger sequencing was the genome of Theobroma cacao, 2010, an economically important tropical fruit tree crop. The genome was sequenced in a consortium, "The International Cocoa Genome Sequencing consortium (ICGS) " and produced a total of 17.6 million 454 single end reads, 8.8 million 454 paired-end reads, 398.0 million Illumina paired-end reads and about 88,000 Sanger BAC reads. First by using genome assembly software, Newbler, an assembly was produced with 25,912 contigs and 4,792 scaffolds from the reads obtained from Roche/454 and Sanger raw data. This had a total length of 326.9 Mb, which represents 76% of the estimated genome size. The Illumina reads were used to complement the 454 assembly, by aligning the short reads on the cocoa genome assembly using the SOAP software. A similar strategy that combined NGS reads and Sanger Sequencing was used for other important plant species like the first published apple genome (Malus domestica), cotton (Gossypium Raimond), draft genome of sweet orange (Citrus sinensis) and the domesticated tomato (Solanum lycopersicum) genome.

Third-generation

With the emergence of third-generation sequencing (TGS) some of the limitations from previous methods of sequencing and assembling plant genomes have started to be addressed. This technology is characterized by the parallel sequencing of single molecules of DNA, that results in sequences up to 54 kbp length (PacBio RS 2). In general, long reads from TGS have relatively high error rates (10% on average) and therefore repeated sequencing of the same DNA fragments is required. The price of such technology is still quite high and therefore is generally used in combination with short reads from NGS. One of the first plant genome that used long-reads from TGS, Pacific Biosciences in combination with short reads from NGS was the genome of Spinach having a genome size estimated at 989 Mb. For this, a 60× coverage of the genome was generated, with 20% of the reads larger than 20 kb. Data were assembled using PacBio's hierarchical genome assembly process (HGAP), and showed that long-read assemblies revealed a 63-fold improvement in contig size over an Illumina-only assembly. Another plant genome that was recently published that used long reads in combination with short reads is the improved assembly of the apple genome. In this project a hybrid approach was used, combining different data types from sequencing technologies. The sequences used came from: PacBio RS II, Illumina paired-end reads (PE) and Illumina mate- pair reads (MP). As a first step an assembly from Illumina paired-end reads was performed using well-known de novo assembly software SOAPdevo. Then using a hybrid assembly pipeline DBG2OLC, the contigs obtained at the first step and the long reads from PacBio were combined. The assembly was then polished with the help of Illumina paired-end reads by mapping them to the contigs using BWA-MEM. By mapping the mate-pair reads on the corrected contigs they scaffold the assembly. Further BioNano (https://bionanogenomics.com/) optical mapping analysis with a total length of 649.7 Mb, were used in the hybrid assembly pipeline together with the scaffolds obtained from the previous step. The resulting scaffolds were anchored to a genetic map constructed from 15,417 single-nucleotide polymorphisms (SNPs) markers. For better understanding of the number and diversity of genes that were identified, ribonucleic acid RNA-seq, were used. The resulted genome has a dimension of 643.2 Mb getting closer to the estimated genome size than the previous published assembly and a smaller number of protein-coding- genes.

The use of long reads in the plant genome assemblies became more popular, for reducing the number of scaffolds and increasing the quality of the genome by improving the assembly and coverage in regions that are not clearly defined by NGS assembly.

GENOMICS OF SALINITY TOLERANCE IN PLANTS

Plants are frequently exposed to wide range of harsh environmental factors, such as drought, salinity, cold, heat, and insect attack. Being sessile in nature, plants have developed different strategies to adapt and grow under rapidly changing environments. These strategies involve rearrangements at the molecular level starting from transcription, regulation of mRNA processing, translation, and protein modification or its turnover. Plants show stress-specific regulation of transcription that affects their transcriptome under stress conditions. The transcriptionally regulated genes have different roles under stress response. Generally, seedling and reproductive stages are more susceptible to stress. Thus, stress response studies during these growth stages reveal novel differentially regulated genes or proteins with important functions in plant stress adaptation. Exploiting the functional genomics and bioinformatics studies paved the way in understanding the relationship between genotype and phenotype of an organism suffering from environmental stress. Future research programs can be focused on the development of transgenic plants with enhanced stress tolerance in field conditions based upon the outcome of genomic approaches and knowing the mystery of nucleotides sequences hidden in cells.

Nature's rage influences plants in the form of various biotic and abiotic stresses. Extreme abiotic stress conditions, such as salinity, flooding, heat, drought, and cold, as well as heavy metal toxicity and oxidative stress affect plants in many different ways. Human activities exacerbate these stress conditions to a greater extent. All the abiotic and biotic stresses, including various pathogens, cause havoc to plants eventually limiting their growth and yield potentials. About 50% of crop yields are reduced due to abiotic stresses, making them the major cause of crop failure worldwide. Abiotic stresses are a serious threat to the sustainability of agricultural industry. Naturally, a number of stresses combine with each other and act together; therefore, the negative effects are aggravated to a greater extent when compared to a single stress factor. To combat these stresses, combinations of diverse pathways are triggered.

In physical terms, stress is defined as a mechanical force per unit area applied to an object. It is difficult to measure the exact force applied by the stresses because the plants are immotile. This makes it harder to define stress in biological terms. A condition, which may act as a stress for one plant, may be ideal for another plant. Hence, a biological stress can most suitably be defined as a harsh condition or force that impedes the normal functioning of a biological system such as plants.

The plasma membrane serves as a barrier that separates a cell from its surrounding environment. Some of the small lipid molecules like steroid hormones are able to pass through this membrane and diffuse into the cytoplasm, whereas the membrane does not allow the water soluble molecules, such as ions, proteins, and other large molecules, to pass through it. Cells start responding when extracellular molecules come in contact with the plasma membrane. This foreign molecule is called an elicitor, and the protein that is present on the cell membrane and interacts with the elicitor is called a receptor. A number of biotic and abiotic stress signals serve as elicitors for the plant cells.

Salinity Stress and its Causes

Total amount of dissolved mineral salts in water and soil is termed as salinity. These salts comprise electrolytes of anions (majorly CO_3^{2-}, SO_4^{2-}, Cl^-, NO_3^-, HCO_3^-) and cations (majorly Ca_2^+, K^+, Mg_2^+, Na^+). Salts that are soluble in water get deposited in the upper layer of soil to a greater extent that hinders the agricultural productivity of that land area. Although fewer salts are present in the rainwater, these salts can be accumulated in the soil over a certain period of time. Salts can also be deposited by soil transported by wind from far off places. Impure irrigation water also contributes to the level of deposited salts in the agricultural lands.

Salinity stress is one of the main abiotic stresses and is considered as a restraint to crop yield. Increased salinization of cultivable land has disastrous effects worldwide. Hyper-osmotic and hyperionic stresses are caused by increased salinity, which can lead to plant death. A number of factors are responsible for causing salinity in a given area such as the extent of precipitation or evaporation and weathering of rocks. Deserts have high salinity due to the fact that the rate of evaporation is greater than the rate of precipitation.

All the key processes within a plant are significantly influenced when the plant is exposed to salt stress. The water stress resulting under salt stress affects the leaf growth and development. Cell division and expansion as well as stomatal opening and closing are negatively influenced by the salinity stress. If the stress condition prevails, then the ionic stress strikes, and eventually, a major decline in photosynthetic rate occurs, and the leaves start to die.

Deforestation is a leading cause of salt stress. Heavy salt–rich irrigation is the major cause of salinity in agricultural lands. The process of evapotranspiration is responsible for the retention of excessive amounts of salt in the soil every year. This is due to abundant loss of water as a result of both evaporation and transpiration. Almost all of the main agricultural crops are sensitive to salt stress that results in serious damage to the yields of the crops. Soil contents altered by the deposition of large amounts of salt in the soil, and as a result, soil becomes less porous reducing soil aeration and water transport. Salinity stress and drought stress are quite similar in terms of physiology.

Stress Signaling Pathways

The receptors present on the plant cell surface receive the stress signals and transfer them downstream, resulting in the production of secondary messengers, e.g. reactive oxidative stress (ROS), calcium, and inositol phosphates. Calcium level is further controlled by these messengers within the cell. As a result of this disturbance in the intracellular Ca^{2+} level, the Ca^{2+} sensors are triggered, which change their conformation in a calcium-dependent manner. These sensors initiate a phosphorylation cascade by interacting with their respective partners and activate the stress responsive genes or the transcription factors that regulate stress response genes. The products of stress response genes help in plant survival and mitigate the stress conditions. Production of hormones (such as ethylene, salicylic acid, and abscisic acid (ABA) takes place because of changes in gene expression under the stress. Initial signal is amplified by messenger 'Sensor' stress response molecules, and a secondary signaling pathway may be induced. Such molecules which do not take part directly in signaling but play a role in alteration of signaling components are called accessory molecules.

The stress responsive genes can be divided into two major categories: early- and late-induced genes. Early-induced genes are prompted immediately after stress signals are received, and most of the times, they express in shorter period. In this category, a number of transcription factors are included because they do not require synthesis of new proteins for their stimulation. In contrast, late-induced genes are expressed slowly under the stress condition, i.e., express after hours of receiving stress signals, and their expression is persistent. In this gene category, major stress responsive genes, such as (COR cold responsive), KIN (cold induced), or RD (responsive to dehydration), and membrane stabilizing proteins, osmolytes, antioxidants, and LEA (late embryogenesis abundant)-like proteins are included.

Salt Tolerance

The percentage of biomass production in saline conditions in comparison with normal growing environment during an extended period is known as salt tolerance. In this regard, vivid variations are found among different plants due to the fact that decline in growth is dependent on the length of time over which plants are growing in the salinity-affected soils. For example, a salt-tolerant plant like sugar beet may undergo 20% decrease in dry weight when grown in 200 mM NaCl. In contrast, a moderately tolerant plant like cotton may undergo 60% decrease in the dry weight, whereas a sensitive plant like soybean may become dead and a halophyte like Suaeda maritima may grow to its full potential in the 200 mM NaCl condition. Evaluation of salt tolerance for perennial plant species can also be done based on survival rate. A marked decline in growth rate is observed in both salt-tolerant as well as nontolerant species during a short time in salt stress. This has been seen in the case of durum wheat and bread wheat, where durum wheat is more salt sensitive and the same was also observed in the case of barley and triticale. This led to realizing the importance of time frame and the mechanisms that different plants use at different growth stages when exposed to salinity.

Mechanisms of Salt Tolerance

The initial discovery of biochemists that enzymes of halophytes and nonhalophytes are equally tolerant to increased levels of NaCl is found to be true. This was explained by the example of enzymes obtained from a halophyte Atriplex spongeosa and those obtained from peas or beans that were equally sensitive to NaCl. This is because most enzymes get inhibited at Na^+ concentration more than 100 mM, and some are observed in the case of Cl^-. Even K+ can also inhibit enzymes when present in 100–200 mM concentrations. Hence, the salt-tolerant mechanisms can be divided into two main categories: (1) preventing or reducing the amount of salt being uptake by plant tissue and (2) reducing the concentration of salt present in the cytoplasm. These both types of mechanisms are found in halophytes, which not only exclude salt very effectively but also quite effectively compartmentalize the excess salt in cell vacuoles. Due to this reason, halophytes are able to grow in saline soils far better and for longer time spans than other plants.

Conventional ways to Manage Salinity

Accumulation of large amounts of salts in the water around the root area is referred to as soil salinity. Plants can tolerate soil salinity by two processes: salt exclusion and salt inclusion. Plants, which are able to eliminate salts from the whole plant or specific plant tissues, are known as salt excluders. Such plants possess low Na^+ and Cl^- content as the membrane permeability prefers K^+ over Na^+ uptake in these plants. On the other hand, salt accumulators can withstand high salt concentrations by two approaches. The first approach is the enduring increased amounts of intercellular salts. The second approach is through the elimination of surplus amounts of salt from the plant because the roots of these plants can absorb salt ions but prevent their harmful effects.

To recover the agricultural lands from salt stress and for increased yields, it is necessary to remove excess amounts of salts from the root region. The common strategies used for this purpose are leaching, scraping, and flushing. As these methods were quite costly, new approaches were introduced for contending salt stress. One of them is the use of halophytes in salinity-affected lands. Halophytes are the plants that can exclude the deposited salts from the soil surface in addition to withstanding high levels of accumulated salts. For this purpose, some halophytes possess salt glands, which are specialized leaf cells having the ability to expel salt. Some others use salt hairs present on stems for this purpose while some have stomatal guard cells, which regulate the rate of transpiration according to the surrounding salt concentration. Another strategy used to protect the plants from the injurious effects of salinity is foliar feeding of nutrients. This enhances plant salt tolerance by relieving plants from Na^+ and Cl^- injury.

Soil salinity can also be controlled using better farm management practices. In this regard, improved irrigation methods, such as drip irrigation, can be used to apply controlled amount of water to the land. In rain-fed areas, crop rotation of annual crops with perennial crops (having deep roots) should be practiced to re-establish the equilibrium between rainfall and used water. This will avert the water tables from rising and delivering salts to the surface.

Genetic Responses to Salinity

Genetic response in case of salinity stress takes in a complex mechanism that is used by plants to up-regulate or down-regulate (increase or decrease) the production of specific gene products (protein or RNA). These mechanisms have been recognized at different stages of central dogma process like from transcriptional initiation to RNA processing, post-transcriptional modification, and translation, and to the post-translational modification of a protein. Understanding about the transcriptional behavior of plants provides a detailed knowledge about the gene expression at mRNA level. Transcriptional profiling is widely used to screen out candidate genes involved in stress responses. Till now, massive information about the salt responsive genes, transcription factors which either up-regulated or down-regulated, has been identified using transcriptome profiling methodology. Further genomic approaches contribute significant role in encoding, cloning, and characterization of these genes. Gene expression under the certain conditions altered by transcription factors. These factors are considered the most important switches that up-regulate or down-regulate the gene expression. Among them, bZIP,

WRKY, AP2, NAC, C2H2 zinc finger gene, MYB and DREB family proteins comprise a large number of stress-responsive members. These transcription factors have the capacity to alter the gene expression by cis-acting specific binding in the promoter region of broad range of genes.

Up-regulation in the expression of bZIP genes were observed in sensitive wheat cultivar under persistent salinity stress and down-regulation in salt-tolerant variety. It predicts the role of NAC transcription factor in salinity tolerance in both rice and wheat cultivars. In rice, transcriptional regulators, such as DREB1/CBF, DREB2, and AREB/ABF, have been demonstrated to play a significant role in abiotic stress responses. Transcription factors, such as OsNAC5 and ZFP179, show an up-regulation under salinity stress, which may regulate the synthesis and accumulation of proline, sugar, and LEA proteins that in turn play an integral role in stress tolerance.

Full-length cDNA is a vital resource for studying the full functional genes in wheat. A group of gene "MYB gene" family analyzed by Zhang et al. that respond to one or more abiotic stress treatments. They isolated 60 full-length cDNA sequences encoding wheat MYB proteins and also construct phylogenetic tree with other wheat, rice, and Arabidopsis MYB proteins to understand their evolutionary relationships and the putative functions of wheat MYB proteins based on Arabidopsis MYB proteins with known functions. Tissue-specific analysis and abiotic stress response expression profiles were also carried out to find potential genes that participate in the stress signal transduction pathway, including the analysis of transgenic Arabidopsis plants expressing the MYB gene, TaMYB32. In Arabidopsis, salt stress results in up-regulation of AtWRKY8 gene expression, which directly binds with the promoter of RD29A. A large number of genes and transcription factors are up-regulated in response to salinity in different plant species, which serve diverse functions, senescence-associated genes. Among stress-responsive genes, the SOS transcription gene family is considered to play a very stimulating role in ion homeostasis, thereby conferring salt tolerance. Most of the salinity responsive genes, such as ROS-scavenging and osmotic-regulating genes, are also up-regulated by salinity in salinity tolerant species. Schmidt et al. observed more than 10 extensively up-regulated genes in the halophyte plant species Spartina alterniflora under salt stress. Most of these genes were related to osmotic regulation process.

Recently, Schmidt et al. characterized root-specific salt-responsive ERF1 (SERF1) transcription factor gene in Oryza sativa that showed a root-specific induction upon salt and H_2O_2 treatment. Plants deficient for SERF1 are more sensitive to salt stress compared with the wild type, although constitutive overexpression of SERF1 improves salinity tolerance. Different types of kinases also regulate the activity of transcription factors and have been found to be significant players of plant adaptation to salinity stress. Serra et al. studied the OsRMC encodes a receptor-like kinase and described as a negative regulator of salt stress responses in rice. Two transcription factors, OsEREBP1 and OsEREBP2, belonging to the AP2/ERF family were shown to bind to the same GCC-like DNA motif in OsRMC promoter and to negatively regulate its gene expression. Basic region/leucine zipper (bZIP) TFs (Transcription factors are proteins involved in the process of converting, or transcribing, DNA into RNA. Transcription factors include a wide number of proteins, excluding RNA polymerase, that initiate and regulate the transcription of genes) possesses a basic region that binds DNA and a leucine zipper dimerization motif. A bZIP class of ABRE-binding transcription factor, known as OSBZ8, has also been identified from rice and has been shown to be highly expressed in salt-tolerant cultivars than in salt-sensitive one. Moreover, OSBZ8 has been shown to be activated/phosphorylated by a SNF-1 group of serine/threonine kinase in the presence of Spd during salinity stress.

Sairam et al. isolated and analyzed the expression response of wheat lip19 (encoding bZIP-type transcription factors) against cold stress. Further analysis confirmed the upregulation of Wlip19 gene in a freezing-tolerant wheat cultivar than in a freezing-sensitive cultivar, while under drought and exogenous ABA application, higher activity of Wlip19 also observed. Heterologous expression of Wlip19 in tobacco has showed a significant increase in abiotic stress tolerance. Alternative splicing of RNA/mRNA played a critical role to cope stress condition especially abiotic stress by switching on/off the transcriptional activities. These splicing factors and spliceosomal proteins mainly involved in plant growth, development process, responses to external environmental factor by affecting the cellular process, cell fate, plant immune/defense system, and tolerance efficiency. All these processes point to critical role of the splicing/alternative splicing under abiotic stress environment.

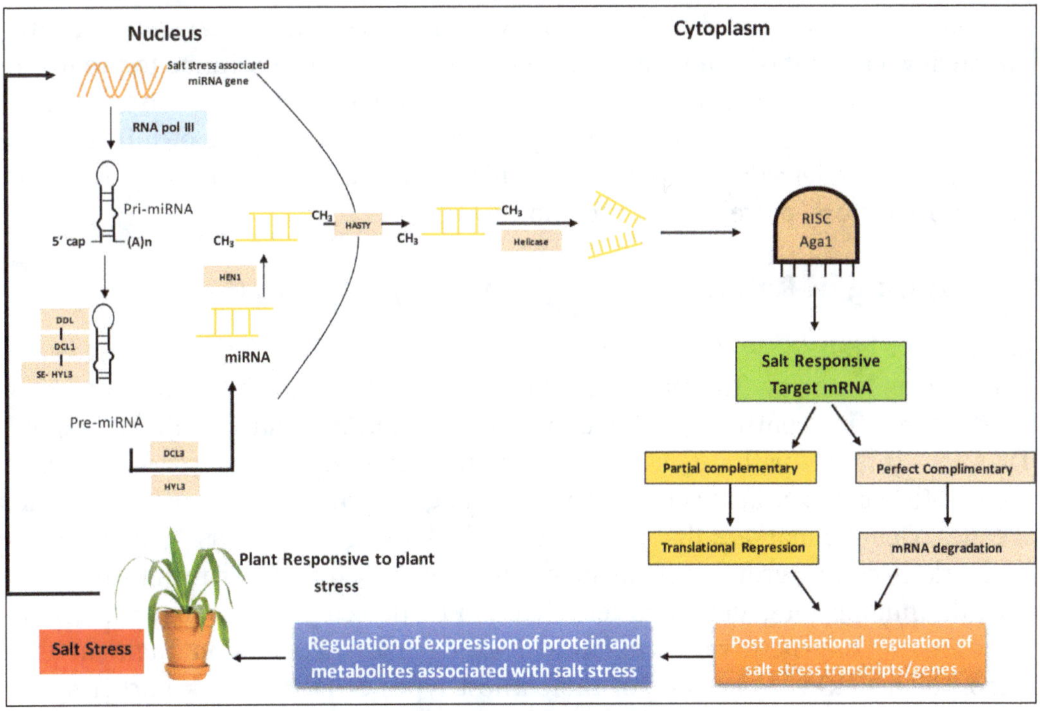

A pathway showing miRNA-mediated post-transcriptional regulation of salt stress–responsive plant genes.

In addition to protein coding genes, recently discovered microRNAs (miRNAs) and endogenous small interfering RNAs (siRNAs) have emerged as important players in plant stress responses. Therefore, post-transcriptional gene regulation plays a crucial role in the plant salt response. Initial clues suggesting that small RNAs are involved in plant stress responses stem from studies showing stress regulation of miRNAs and endogenous siRNAs, as well as from target predictions for some miRNAs. There has been strong evidence leading to the proposal that miRNAs are hypersensitive to abiotic or biotic stress as well as to diverse physiological processes. Drought, cold, and salinity are major abiotic stresses for plants; all of these conditions strongly induced miR402 overexpression. Numerous studies on plants, such as Arabidopsis thaliana and Oryza sativa, have been studied with respect to miRNA expression analysis and have revealed an important role for miRNAs in response to abiotic stress.

Various studies with respect to miRNAs profiling under abiotic stress point out the several differentially expressed miRNAs. In response to salt stress, miRNAs, such as miR396, miR394, miR393, miR319, miR171 miR169, miR168, and miR167, were up-regulated; whereas the miR398 was down-regulated in Arabidopsis, thus indicate a role for miRNAs in the response to salt stress.

Up-regulation of miRS1 and miR159.2 in response to salt stress was observed in Phaseolous vulgaris. The expression of miR530a, miR1445, miR1446a-e, miR1447, and miR171l-n was increased, whereas the expression of miR482.2 and miR1450 was decreased during salt stress in Populus trichocarpa. Furthermore, two members of miR169 family namely miR169g and miR169n showed enhanced expression during salinity. With the development of genomics tools and computational algorithms to predict and identify the miRNAs in various plant species, the number of miRNAs associated with salt stress response is increasing. A comprehensive understanding of miRNA-based gene regulation under salt stress will definitely help in elucidating the complex network of regulatory factors, proteins, and metabolites.

QTL Mapping in Relation to Plant Salinity Tolerance

A quantitative trait locus (QTL) is a section of DNA (the locus) that correlates with variation in a phenotype (the quantitative trait). The QTL typically is linked to, or contains, the genes that control that phenotype. Genetic marker is an identifiable fragment of DNA that is linked with a specific point and indicates genetic differences within the genome. Molecular markers act as a 'signs' or 'flags' should not be considered as normal genes. Genetic markers that are tightly linked are referred to as "gene tags". The DNA markers can be grouped in various categories based on their technical requirements, the number of genetic markers that can be detected throughout the genome, and the amount of genetic variation found at each marker. Restriction fragment length polymorphisms (RFLPs) are one of the earliest types of DNA-based marker system, which detect the variation in restriction fragment length by Southern hybridization, which cause single base changes that led to the creation or removal of a restriction endonuclease recognition site to detect shift in fragment size. Although this technique is an important tool in breeding programs, it has been superseded by microsatellite or simple sequence repeat (SSR) markers and is now rarely used. SSR markers detect variation in the number of short repeat sequences, usually two or three base repeats that allow the detection of multiple alleles. The expressed sequence tag (EST) databases have now opened the opportunity for the identification of single nucleotide polymorphisms (SNPs) that occur at varying frequencies depending on the species and genome region being considered.

Table: QTLs for 'Salt Tolerance' (ST) in various plants identified by SSR markers.

Crop Plants	Locus	Traits
Wheat (Triticum aestivum L.)	Kna1 Nax1	Controls the selectivity of Na+ and K+ transport from root to shoot and maintains high K^+/Na^+ ratio. Both are involved in decreasing Na+ uptake and enhancing K^+ loading into the xylem.
Rice (Oryza sativa L.)	qRL-7, qDWRO-9a and qDWRO-9b qBI-1aand qBI-1b QNa, QNa:K, SKC1/OsHKT8 qDM-3 and qDM-8, qSTR-6 qNAK-2 and qNAK-6 Saltol Saltol and nonSaltol QKr1.2	Play important roles in root length and root dry weight at seedling stage under saline conditions Regulate K^+/Na^+ homoeostasis Improve Na+/K+ ratio under saline conditions. Improve Na^+/K^+ ratio Controls shoot Na^+/K^+ homoeostasis Control shoot Na^+/K^+ homoeostasis Controls K^+ content in root.
Barley (Hordeum vulgare)	Five QTL for ST were identified on chromosomes 1H, 2H, 5H, 6H, and 7H, which accounted for more than 50% of the phenotypic variation. A locus HvNax3 on the short arm of chromosome 7H in wild barley (Hordeum vulgare ssp. spontaneum) accession CPI-71284-48	Enhance vegetative growth under saline stress. Reduces shoot Na^+ content by 10–25% in plants grown under salt stress (150 mM NaCl).
White clover (Trifolium repens L.)	Several QTLs for ST, some at common locations, but each of low scale.	Affect ST during vegetative stage.
Soybean (Glycine max (L.) Merr.)	A major QTL for ST was identified near the Sat091 SSR marker on linkage group (LG) N. Eight QTLs for ST were detected. A major QTL for ST was detected.	Maintains growth under salt stress. Maintains growth under salt stress. Maintains healthy growth under salt stress.

These DNA marker types could be associated with quantitative traits, which are known as quantitative trait loci (QTLs). Mapping of QTLs for salt tolerance have been a slow process due to the complexity of this trait and poor understanding about it. Ren et al. discovered a gene locus named as QTL SKC1, which codes for a transporter that removes Na^+ from the xylem. Several QTLs have been identified in different crop plants. QTLs for yield and physiological characteristics were identified at a late stage of growth of barley under salinity stress. A total of 10 traits were considered for which 30 QTLs were identified under salt stress and nonstress conditions. Of these 30, 13 QTLs were discovered under salt stress. In white clover, QTLs for salt tolerance were identified at the vegetative stage of plants and the results showed that, in white clover, multiple QTLs are responsible for controlling the salt tolerance. However, QTLs for salt tolerance in tomato were detected at the seedling stage of Solanum pennellii and Solanum lycopersicoides plants. In S. pennellii, four major QTLs were detected, for salt tolerance, on chromosomes 6, 7, and 11, whereas in Solanum lycopersicoides six major QTLs were identified under salt stress on chromosomes 4, 6, 9, and 12. QTLs for salt tolerance in soybean were identified on chromosome 3.

GENOME REFERENCE SEQUENCES AND RE-SEQUENCING

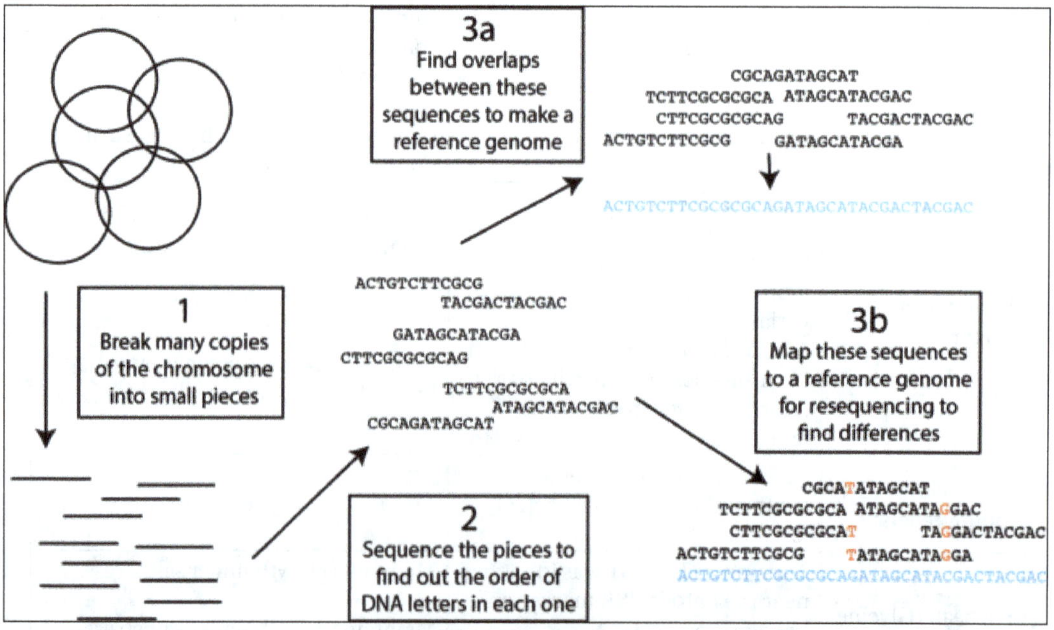

Assembly and re-sequencing.

When sequencing an entirely new species, there is no pre-existing, similar sequence to help us out in determining where genes are located. The goal is to produce what is known as a 'reference genome sequence'. To produce a reference sequence, we use a process termed assembly. Assembly is the process of putting together individual sequencing reads, much like a puzzle, to make longer stretches of sequence called contigs.

The best possible result is to end up with a single sequence that represents the entire bacterial chromosome. In reality, we cannot always complete the puzzle and there are pieces that we cannot put together. Contigs are assembled using computer programs that look for overlaps between sequencing reads.

Genome Annotation

Once a reference assembly has been produced, the next step is to find the genes, a step known as genome annotation. There are computer programs designed to do this. They find parts of the genome sequence which look like they might encode protein sequences. This is important because it identifies the functional toolkit of the bacterium. It also helps us understand differences between bacterial genomes. When we look at changes in the genome that might be related to antimicrobial resistance, for example, we can identify which genes are involved.

Different Sequencing Technologies have Different Roles

Over time, more genome sequences have become available for different species. Developments in genome sequencing technology have also made sequencing easier and cheaper. Some new genome sequencing technologies allow us to improve reference sequences because they have much longer sequencing reads (e.g. PacBio and Oxford Nanopore can produce 10,000-50,000 bases), which help to join contigs together and produce more complete reference sequences. Other technologies produce relatively short reads (Illumina produces 75-150 bases) but generate a very large number of them at low cost. The ability to produce lots of short reads has allowed us to identify similarities and differences between bacteria of the same species cheaply and efficiently (resequencing). This has been crucial in helping us track the spread of bacterial disease.

Resequencing

Sequencing a new example of a species for which a reference genome already exists is known as resequencing. Discovering what makes the new bacterial genome different from the reference is much easier than assembling a new reference genome. Instead of assembling a new genome sequence, we use read mapping. For each sequencing read from the new bacterium, we look for the most similar part of the reference genome and place it there. We can then look for differences between the mapped reads and reference genome, which represent mutations in the genome of the new bacterium.

SEQUENCING STRATEGIES

Genome sequencing, assembling and annotation have been major priorities in plant genetics research during the past 20 years. The release of draft reference genomes have typically constituted major milestones and have proven to be invaluable for the analysis and characterization of genome architecture, genes and their expression, diversity and evolution. The expansion of sequence information in a growing number of taxa has contributed to comparative studies and the implementation of molecular breeding and biotechnology approaches for crop improvement. The construction of the first plant genomes was made possible by applying considerable resources, coordination and effort to enabling automated Sanger-based sequencing technologies and computational algorithms. Starting in 2005, a series of technological revolutions in DNA sequencing, driven in large part by the goal of affordable personalized genome sequencing, radically changed the sequencing model. First, new technologies drastically increased throughput while reducing costs and times in data collection. Additional technologies then enabled long single-molecule reads and algorithms that were more suitable to resolve complex genomes.

In addition to these advances, the genomics community has benefited from the development and implementation of complementary mapping technologies and methods that have facilitated the scaffolding of sequences and integration to genetic maps.

The Development of Sanger Sequencing for De Novo Assembly of Plant Genomes

The construction of reference genomes was initially enabled by technological advances in sequencing using the Sanger method. During the 1980s and 1990s, the introduction of thermal cycle sequencing, single-tube reactions and fluorescence-tagged terminator chemistry facilitated the development of high-capacity sequencing platforms. Additional improvements in parallelization, base quality assessment, read length and cost-effectiveness were achieved by the development of automatic base-calling and capillary electrophoresis. With no major modifications made in the past years, automated high-throughput Sanger sequencing is performed by parallel reactions that include a mixture of the DNA template, primer, DNA polymerase, and deoxynucleotides (dNTPs). A proportion of dideoxynucleotide terminators (ddNTP) are included in the reaction, each labelled with a different fluorescent dye. DNA molecules are extended from templates using a thermal cycling reaction and terminated by random incorporation of the labelled ddNTPs, which are detected by laser excitation of the fluorescent labels after capillary-based electrophoresis. The differences in dye excitation profiles are recorded and translated by a computer to generate the sequence. Primary analysis software then calls nucleotides from the raw sequences, assigning a corresponding quality score at each position.

The complete sequencing of the first bacterial genomes as well as the creation of initiatives aimed at sequencing the genomes of Sacharomyces cerevisae, Caenorhabditis elegans, Drosophila melanogaster and Homo sapiens provided the technical and technological framework for the initial sequencing of genomes in plants. These projects validated the idea of applying a scaled-up form of shotgun sequencing. Shotgun sequencing relied on computer algorithms to enable in silico assembly of overlapping sequencing reads derived from randomly-generated subclones. The development of software suites such as Phred, Phrap and Consed allowed calling bases, setting individual base quality, assembling overlapping reads, assigning assembly quality scores, viewing final assemblies and extracting consensus sequences. Two major genomic shotgun sequencing strategies were defined at that time: (1) whole-genome shotgun sequencing (WGS) and (2) clone-by-clone, also referred to as BAC-by-BAC sequencing. In WGS, genomic DNA is randomly sheared and the ends of the cloned fragments are directly sequenced and assembled. This strategy is the simplest, and it was initially used in small bacterial and yeast genomes. Later, it was also used in D. melanogaster and one of two initiatives aimed at sequencing and assembling the reference human genome. Major improvements to de novo WGS assembly came from using strategies that relied on paired-end reads from multiple libraries with different average insert sizes and the optimization of software with algorithms that use end-sequence distance information from these libraries.

The second Sanger sequence assembly strategy, clone-by-clone, was successfully deployed in projects aimed at complex eukaryotic genomes. In clone-by-clone genome assembly, shotgun sequencing is performed in libraries derived from individual genomic large-insert clones, selected in a minimum tile path according to physical and genetic map information. The most common type of large-insert clone is the bacterial artificial chromosome (BAC), which can stably carry genomic inserts ranging from 100 to 300 kb and is relatively easy to maintain and purify. Accordingly, this method is usually referred to as BAC-by-BAC, although additional vector systems have been used in assembly projects, including yeast artificial chromosomes (YACs), P1 artificial chromosomes (PAC), transformation-competent artificial chromosomes (TACs), cosmids and fosmids. The two major genomic shotgun-sequencing approaches, WGS and BAC-by-BAC, had advantages and disadvantages when applied to Sanger-based sequencing platforms, depending on the genome of interest. The clone-by-clone approach benefited from working in small units, effectively reducing complexity and computational requirements. This approach minimized problems associated with the misassembly of highly repetitive DNA and therefore provided a better, more complete assembly in plants and other complex eukaryotic genomes. WGS projects were computationally intensive and were less effective bridging repetitive regions in complex genomes but benefited from considerably lower cost, time and logistics.

The first completed reference plant genome was from the model system Arabidopsis thaliana, accession Columbia. At that time, it was only the third multicellular eukaryotic genome to be published, after C. elegans and D. melanogaster. The nuclear genome of Arabidopsis is distributed in five chromosomes, and it is only approximately 4% the size of the human genome. The A. thaliana genome initiative used multiple types of available large-insert libraries including cosmids, BACs, PACs and TACs. Shotgun clones were constructed and then mapped by restriction fragment fingerprinting as well as screening with hybridization or polymerase chain reaction (PCR) markers. End sequences for 47,788 BAC clones were further used to anchor clones, integrate contigs and help select a minimum tiling path. Each of 1,569 clones in a minimum tiling path were selected, sequenced bidirectionally and assembled at estimated error rates of less than 1 in 10,000 bases. Direct PCR products were used to close some gaps and YACs allowed the characterization of telomere sequences. As initially published, the total length of sequenced regions was 115.4 Mb, in addition to an estimated 10 Mb nonsequenced centromeric and rDNA repeat regions. Since the original publication, the Arabidopsis genome sequence reference has been subjected to several rounds of improvements, each time reducing gaps and extending the sequence towards the centromeric regions.

The second published plant genome reference was rice (Oryza sativa). While the rice genome is more than 2-fold the size of Arabidopsis, approximately 390 Mb, it is one of the smallest genomes of any major crop, less than 15% the size of the human genome. Like Arabidopsis, the rice genome was completed using a Sanger-only clone-by-clone approach that required the initial construction, fingerprinting and physical mapping of a large number of random BACs and PACs. In total, 3,401 mapped clones in a minimum tiling path were selected from the physical map, randomly sheared and individually end-sequenced to approximately 10-fold coverage. Clone sequences were assembled and low-quality regions were finished using targeted sequencing. Gaps were closed and low-quality regions resolved by sequencing PCR fragments, plasmids and fosmids.

The draft reference genome of Maize, one of the most important crops in the world, is considered the last major published plant genome project based primarily on a Sanger BAC-by-BAC strategy. At 2.3 Gb and spanning 10 chromosomes, the nuclear genome of maize is considerably larger than that of rice and Arabidopsis, approximately 3/4 the size of the human genome. A set of 16,848 minimally overlapping BAC clones, derived from an integrated physical and genetic map, were selected and end-sequenced. The assembly was performed after adding additional data derived from cDNA sequences and sequences from subtractive libraries with methyl-filtered DNA and high C_0t techniques, resulting in a whole-genome assembly (B73 RefGen_v1) made of 2,048 Mb in 125,325 sequence contigs and 61,161 scaffolds. Unlike the completed genomes of rice and Arabidopsis, most sequenced BACs in the first version of the maize draft genome are unfinished. Gaps and low-quality regions in BACs were not systematically closed by PCR sequencing or other target approaches. Therefore, while the BACs used in the minimum tiling path were mapped, the order and orientation of individual contigs within a single BAC could be incorrect. Subsequent versions of the genome have been improved by targeting gaps and adding alternative sequencing strategies.

Finally, it is important to mention that a significant number of plant genome sequencing initiatives have used WGS strategies, which provide a considerable reduction in time and cost associated with cloning, construction, mapping and selection. Sanger WGS genome projects included those of poplar tree, grape, and papaya. Later refinements to the process enabled the sequencing of Brachypodium distachyon as well as the larger genomes of Sorghum bicolor (730 Mb) and soybean, an ancestral tetraploid (1.1 Mb). It should be noticed that, as demonstrated by the Maize genome project, the two Sanger shotgun assembly approaches, as well as later sequence technologies, are not mutually exclusive and may be complementary to increase quality and coverage.

The high cost and logistics of plant projects based on clone-by-clone Sanger sequencing required extensive funding, the creation of large collaborative consortia and several years of fingerprinting and sequencing work. The cost of the project by the Arabidopsis Genome Initiative has been estimated at US$70 million. The International Rice Genome Sequencing Project (IRGSP), which included groups from 11 different nations, took over 5 years to complete. During its early stages, IRGSP had estimated that the project would take 10 years and cost a staggering US$200 million. The Maize draft genome was accomplished by multiple laboratories at an estimated cost of tens of millions in a joint NSF/DOE/USDA program. It is worth noticing that, while the cost and time required to accomplish Sanger WGS projects are in fact lower than those based on a clone-by-clone approach, they are still considerable for today's standards. The sequencing of the 1.1-Gb soybean genome, the largest published plant genome based on a Sanger WGS approach, provides an example of such a cost. It was completed in less than two years although it took a group of 18 institutions several million dollars to generate and assemble more than 15 million Sanger reads from multiple libraries with average sizes ranging from 3.3 kb to 135 kb.

Besides cost and time considerations, these early Sanger-only projects posed considerable technical challenges. Despite the extensive resources deployed towards the sequencing of the Arabidopsis and rice genomes, which are usually considered as finished, as well as other projects mentioned in this review, they all have representation gaps. A considerable number of gaps correspond to regions that are "unclonable" under the conditions used to prepare BAC and other genomic libraries. Although many of these regions correspond to tandem repeats such as telomeric sequences and other repetitive regions, it may also include gene space. Moreover, the maximum length of quality Sanger reads, usually 800–900 bp, as well as technical issues associated with the sequencing of DNA stretches with strong secondary structures or extensive homopolymers, create conditions for additional sequencing gaps, even in regions with physical coverage.

Finally, most plant genomes are characterized by elevated proportions of highly repetitive DNA and by the presence of segmental duplications or full genome duplications due to polyploidization events, which can be problematic during assembly. The 1C genome content in Maize, for example, is smaller than in humans but it consists of higher proportions and larger tracks of high-copy elements such as retrotransposable elements. At least some of the differences between the assembled and estimated genomes of the Maize line B73 could be attributed to the assembly-based collapse of highly similar long terminal repeats (LTRs) at the end of retrotransposons. It is important to note that all the Sanger-only initiatives corresponded to plant species with genomes that were considerably smaller than the average 5.8-Gb plant genome. Plant genomes have a considerably wider size range than in mammals, and in some important crops (e.g. wheat), nuclear genomes can be more than 15 Gb long, well beyond the practical realm of Sanger sequencing. Although BAC-by-BAC approaches can reduce complexity by more than 10,000 fold, Sanger-based assembly remains difficult and prohibitively expensive in plant genomes of moderate or large size. The WGS approach is even more sensitive to the complexity of plant genomes as it increases the potential for assembly artefacts due to haplotype and homeolog collapse in regions with high identity. Reductions in time and cost in WGS projects are achieved at the expense of assembly fidelity in repetitive regions and expanded need for computational resources.

Next-generation Sequencing Technologies Applied to De Novo Assembly of Plant Genomes

Second-generation Sequencing Technologies

As indicated above, successful whole-genome sequencing projects have been achieved with the use of Sanger technology. However, such projects require dealing with several complicating factors, including high costs and relatively long turnaround times to completion. The emergence of next-generation sequencing (NGS) technologies has changed this paradigm, both by reducing costs and increasing sequencing throughputs, while at the same time introducing complexity related to the relative short reads of NGS reads. Several NGS technologies have emerged in the past 7 to 8 years. All follow a relatively uniform approach to library construction and sequencing. To complete sequencing:

- Universal adapters are ligated at the end of single DNA molecule templates.
- Adapter-ligated DNA templates are amplified via PCR to create a cluster of identical isoforms.
- Clusters are loaded on sequencers and nucleotide incorporations occur in parallel on millions of clusters.

These generate an amplified signal that is recognized by the platform and translated into a base call.

The most widely used NGS technology nowadays is the one commercialized by Illumina, whose high-throughput instrument, the HiSeq4000, can produce up to 1.5 Tb of sequencing data in approximately 3.5 days. In the Illumina sequencing platform, sequencing templates generated during library construction are immobilized on a solid surface, and a "bridge PCR" approach allows for the localized amplification of millions of single DNA molecules, thus generating millions of clusters, each containing thousands of copies of the original DNA molecules. Sequencing then is performed using a sequencing-by-synthesis approach where single-base extension allows the incorporation of a fluorescently labelled nucleotide (a blocking chemical moiety at the 3' hydroxyl end allows the incorporation of one base only). Once incorporated, the label is detected and the resulting signal subsequently translated into a base call. Finally, the fluorescent dye and the blocking 3' agent are cleaved, allowing the next single base incorporation event to occur. Through the use of alternating cycles of base incorporation, image capture and dye cleavage, the Illumina sequencing technology can produce reads that are up to 300 bp in length. The relatively high error rate (0.1% or 10 times higher than Sanger sequencing) can be compensated by very high sequencing coverage, thus allowing random errors at any given base position to be ignored below a certain frequency threshold. The relative short read of Illumina sequencing reads can be explained by several noise factors accumulating after each cycle, including phasing, where imperfect single-base incorporation and imperfect cleavage of the dye and 3' hydroxyl blocking moiety lead to the accumulation of copies of various lengths within a cluster, and the subsequent increase of signal-to-noise ratio after each cycle.

Third-generation Sequencing Technologies

De novo assemblies of plant genomes have been performed with NGS reads only, either with reads generated on the Illumina platform alone or with reads generated with the Illumina platform combined with reads generated on the Roche 454 second-generation sequencing platform. However, those assemblies generally are fragmented, resulting in low N50 values and a high number of contigs, mostly because of the overall short read length, the complexity of the genome and the presence of conserved regions whose length exceeds the length of NGS reads and thus cannot be extended during the de novo assembly process. The emergence of third-generation sequencing technologies has started to address some of the inherent limitations of sequencing and assembling large and complex plant genomes. Those technologies are characterized by the parallel sequencing of single molecules of DNA (rather than "clusters"), thus avoiding phasing issues, and the resulting sequences tend to be in the kb range, offering the opportunity to assemble genomes and generating longer contigs by encompassing complex and conserved genomic regions and allowing relatively high-confidence assemblies of overlapping reads. However, single sequencing reads tend to exhibit relatively high error rates (15%–25% on average). Deep sequencing coverage or repeated sequencing of the same DNA fragments therefore are required to offset the presence of a high number of sequencing errors. As of today, two companies have developed and commercialized third-generation sequencing technologies, namely, Pacific Biosciences (e.g., 50) and Oxford Nanopore Technologies (e.g., 51). Each company uses vastly different approaches to sequencing. The Pacific Biosciences (PacBio) RS II system uses a sequencing-by-synthesis approach to offer up to 40-kb reads, where base incorporation is monitored in a real-time fashion. Nanoscale holes, described as Zero Mode Waveguides ("ZMW") are located on a chip, where individual polymerases are covalently attached to the surface of each ZMW. Individual nucleotides with a fluorescent label attached to the phosphate chain are incorporated to the elongating strand and the excited dye emits a signal that is captured before diffusion of the released pyrophosphate, and translated into a specific base call. DNA fragments used as template are ligated to "bell-shaped" adapters at both ends, thus facilitating the sequencing of DNA fragments through multiple passes and the creation of a more accurate consensus sequence. The overall stability and activity of the polymerase remain limited by photo damage and the progressive dissociation of the polymerase/template complex from the surface of the ZMW. It is therefore expected that reads generated from smaller DNA fragments will exhibit higher consensus accuracy than reads from larger DNA fragments. Oxford Nanopore Technologies released the MinION sequencing device in early access mode in 2014. Like the PacBio RS II system, the MinION delivers long reads in a real-time fashion, from single molecules of DNA. In that particular case, however, sequencing is performed by measuring the change in ionic current when a single DNA strand translocates through a protein nanopore located in an insulated membrane. The resulting signal is measured and translated into a base call. Because no enzyme is involved in the DNA sequencing process, it is expected that read length will be driven mostly by the physical length of the DNA strand being sequenced. Library construction involves the ligation of two types of adapters to DNA fragment, one "Y-shaped" adapter with a bound protein that unwinds the double-stranded DNA and facilitates the translocation of a single strand through the pore, and one "bell-shaped" adapter at the other end that allows the translocation, and sequencing, of both the sense and antisense strands. Sequencing reads then are generated by aligning base calls from the two strands and producing a higher quality consensus sequence.

Challenges in Assembling Plant Genomes

De novo assembly of genomes has closely mimicked the trends and improvements in sequencing technologies and accompanying sequencing assembly software over the years. The emergence of next-generation sequencing technologies has allowed a much larger number of plant genomes to be sequenced and assembled than what would have been deemed possible with Sanger sequencing alone, mostly because of the costs and labor involved in such projects. However, the complexity of the majority of those genomes still makes it a challenge to resolve them with short reads alone. As a result, most plant genome assemblies are highly fragmented, with large number of contigs and conserved regions of the genome in an unfinished state. The presence of highly conserved repeats often exceeding 10 kb in length represents a major challenge in assembling plant genomes. The most common types of repeats in plants are type II long-terminal repeat (LTR) retrotransposons and their proliferation within a genome often explains most of the structural variations between strains. Their movement also results in genome expansion, where repeats represent, in some instances, more than 80–90% of the structural content of a particular genome. Repeat expansion also can lead to very large genome sizes. While NGS technologies can generate enough raw data to cover an entire genome in a relatively cost-effective manner, assembling such a large amount of data often represents a major computational challenge. For example, the assembly of the loblolly pine genome (22 Gb), which represents the largest genome assembled to date, could be solved only using condensed sets and read pooling prior to assembly. Assembling large and repeat-rich genomes can also be facilitated by using supplemental layers of information, such as the physical distance between "paired" reads (end-sequences generated at both ends of a particular DNA fragment) in mate-pair libraries. Another challenge for de novo assembly of plant genome is the issue of polyploidy. Polyploidy is an important force in plant genome evolution and it is estimated that 80% of all living plants are polyploids, while close to 100% of all plant lineages have a paleo-polyploidy event in their history. As a consequence, some plants species, including economically important crop species like soybean, have entire gene families consisting of highly similar paralogs. Those gene families are the direct result of paleo-polyploidization events where the merger of genomes has been followed by extensive structural rearrangements, including gene loss, and the modification of gene expression for paralogs within a particular gene family. The diploid genomes of progenitor species can be used to determine the origin and structure of contigs when assembling large polyploid genomes. Finally, heterozygosity may represent another important challenge when assembling plant genomes. Outcrossing species like grape, for instance, exhibit up to 13% sequence divergence between alleles, and the existence of such variation will impact contig assembly when both alleles are sequenced in a whole-genome assembly project.

Examples of Plant Genome Assemblies

According to Michael and Van Buren, over 100 plants genomes have been sequenced since 2000, out of which 63% are genomes from various crop species. As indicated above, different Sanger sequencing strategies have been applied with varying degrees of success on several plant genomes. However, the most successful Sanger-based genome assemblies have been obtained from relatively small genomes (Arabidopsis, rice), while de novo assemblies for larger and complex genomes, such as maize, remains partial and unfinished (manual improvements of the maize genome were limited to nonrepetitive regions only). In addition, due to the high costs and labor associated with such approaches, and the need for (in most cases) an international consortium to complete such projects, a vast majority of the most recent genomes have been sequenced using either a hybrid approach, complementing Sanger sequencing with NGS data, or using NGS data alone, from various NGS platforms. Such platforms include Illumina, 454/Roche, and more recently, Pacific Biosciences.

The domesticated tomato genome represents an example of Sanger/NGS hybrid genome assembly. A total of 30,800 BAC clones from three different BAC libraries were shotgun-sequenced and end-sequenced, generating a total of 3.3 Gb of Sanger reads. In addition, 454/Roche shotgun and mate-pair sequencing was performed, both on BAC pools and whole-genome DNA preparation, using different insert sizes and generating a total of 21 Gb of NGS data. The de novo assembly of Sanger and 454 data was performed using the Newbler assembly software and other sequence assembly and alignment tools. Further scaffolding and polishing of the assembly were performed when integrating BAC end-sequence data and additional high-coverage Illumina and ABI/SOLiD data. Taken together, the de novo assembly resulted in 3,761 scaffolds totalling to 782 Mb, with 95% of the assembled scaffold sequences present in 225 scaffolds. The predicted tomato genome size is approximately 900 Mb. The correctness and integrity of the assembly were validated through different means including the alignment of clone end-sequences, publicly available tomato EST sequences, and alignment of BAC contigs from a sequence-based physical BAC map. Interestingly comparison of the tomato, potato and grape genomes supported the existence of two successive whole-genome triplication events in common ancestors that added new gene family members that mediate important fruit functions, such as enzymes involved in ethylene biosynthesis (examples of whole genome duplication or triplication events abound among plant genomes that have been sequenced to date).

Because of the relatively cheap costs involved, a large number of plant genomes have been sequenced and assembled using NGS technologies alone. This includes the assembly of the complex tetraploid genome of cultivated cotton (Gossypium arboreum). The tetraploid cultivated cotton genome has a genome size of approximately 1.7 Gb and is thought to have appeared 1–2 million years ago through interspecific hybridization between diploid A (Gossypium arboretum) and D (Gossypium raimondii) subgenome progenitors. A total of 371.5 Gb of shotgun Illumina data was generated with various insert sizes ranging from 180 bp to 40 kb and complemented with 33,454 BAC end sequences. The assembly was performed with SOAPdenovo, which resulted in 40,381 contigs, anchored and oriented in 7,914 scaffolds, ranging in length from 140 kb to 5.9 Mb with 90% of the contigs included in 3,740 scaffolds.

An example of a smaller, relatively less complex genome assembly is that of the crop species Brassica rapa. An estimated 72× sequencing coverage of the genome was generated, corresponding to Illumina shotgun paired-end data from NGS libraries with insert sizes ranging from 200 bp to 10 kb, and assembled using SOAPdenovo. The resulting assembly was made of 14,207 contigs larger than 2 kb, further assembled into 794 scaffolds, totalling approximately 283.8 Mb and estimated to cover more than 98% of the gene space, based on alignments of 214,425 B. rapa public EST sequences and 52,712 unigenes from the BrGP database. Further assessment of the integrity of the assembly was performed by aligning BAC clone Sanger sequences reported in previous studies.

While a large number of genomes have been sequenced with NGS technologies alone, the relatively short reads of the major NGS platforms that have been used in those assembly projects, combined with the general complexity of most of those genomes, generally require the use of alternative methods to facilitate the assembly or confirm its integrity. These methods rely on the use of various types of NGS libraries, such as mate-pair large inserts, or the use of Sanger-derived sequencing data such as EST or BAC-based shotgun reads. However, scaffolding of NGS contigs, based on using pairing information between NGS reads originating from the same DNA fragment, generally leads to unresolved gaps between contigs, often due to the presence of large repeat regions whose size exceed the length and resolution of short NGS reads. As a result, significant portions of any given scaffold contain large batches of unknown sequences, and of unknown length. To address these issues and improve plant genome assemblies, researchers have developed a series of multifaceted solutions, combining alignment to known public data, such as ESTs or BAC ends, or, when available, reference genomes from related species, integration of physical and genetic map data, or new technologies.

Complementary Approaches to De Novo Assembly of Plant Genomes

Long-read Assembly

NGS assembly strategies based on the use of short reads cannot solve long and identical transposable elements abundantly present in most plant genomes. The use of long reads is expected to address some of those shortcomings and improve the overall quality of de novo assembly by ordering contigs, closing gaps, and improving scaffolding. As a consequence, researchers have started to adopt the single-molecule long-read sequencing technology from Pacific Biosciences in plant genome assembling projects. Spinach is an example of such genome assembly efforts. Spinach is a diploid species with a genome size estimated at 989 Mb. Van Deynze et al. sequenced and assembled the Spinach genome using large fragment libraries of Pacific Biosciences sequence reads. They generated 60 × coverage of the genome, with 20% of the reads larger than 20 kb. Data were assembled using PacBio's Hierarchical Genome Assembly Process (HGAP), which showed that long-read assemblies exhibited a 63-fold improvement in contig size over an Illumina-only assembly, derived from multiple Illumina libraries.

A distinct strategy to long-read assembly, namely, the Illumina TruSeq Synthetic Long-Read (SLR) strategy, is also expected to improve the quality of assemblies generated with short reads only. In SLR libraries, genomic DNA is fragmented to ~10 kb and individual indexed Illumina libraries are generated in parallel from highly diluted pools of sheared DNA fragments. After Illumina sequencing and data deconvolution, the original 10 kb fragments can be reassembled, effectively reducing the complexity level of the assembly and generating very-high quality synthetic long reads that can subsequently be assembled together or used for haplotype resolution.

The use of long reads in de novo assembly is bound to become more prevalent in the near future, reducing the number of scaffolds while at the same time increasing their average length. The use of PacBio in smaller genomes, such as microbial genomes, has already demonstrated that the assemblies often result in contigs corresponding in most cases to individual chromosomes or plasmids present in the microbial cells. Likewise, it is likely that future plant studies will include such long reads, either alone or in combination with short-read NGS data to improve assembly and coverage in questionable regions, and to confirm the integrity of the assembly in a manner similar to Sanger data with current NGS assemblies.

Genetic Anchoring

The emergence of NGS technologies has rapidly led researchers to develop methods and assays for variant discovery in various plant genomes. Some studies have shown that Single nucleotide polymorphisms (SNPs) can be discovered in parental inbred lines using next-generation sequencing. Entire mapping populations also have been simultaneously sequenced and genotyped, in a process known as "genotyping-by-sequencing" (GBS), discovering in the process extensive lists of segregating markers within the mapped population, that can be completed by using known reference maps or sequences to impute missing marker data from individual haplotypes. Various reduced-representation methods have been employed for NGS-derived SNP discovery in plant species where whole-genome shotgun sequencing still remains too expensive for sequencing more than a few individuals. These methods include the use of restriction enzyme digestion–based assays with methyl-sensitive restriction endonucleases, or methods based on sequence capture approaches, to sequence and map gene-rich portions of a genome, and allowing the anchoring of SNPs in a relatively unambiguous manner.

More recently, ultradense linkage maps have been created from genotyping by whole genome sequencing of a genetic mapping population. It has been used to place whole-genome sequencing contigs into a map, thus anchoring, and ordering, sequencing of contigs. Such an approach requires using a genetic linkage map as a framework, into which SNPs derived from the whole genome sequencing assembly can be integrated into a genetic framework derived from low coverage whole-genome sequencing data from a segregating population. The genetic position of the sequence-derived SNPs can then be used to assign chromosomal locations to the contigs harboring them. Such an approach has been used in the context of a whole-genome assembly project in barley where genetic anchoring was applied to a whole-genome assembly. SNPs discovered by sequencing individuals from two mapping populations at low coverage (1×) were placed into genetic maps that had been previously constructed through different means, including SNP array data and GBS, or made from the whole-genome shotgun sequencing data of the population. Their genetic positions then were used to assign chromosomal locations, and integrate into the combined physical and genetic genome framework, approximately two-thirds of all whole-genome shotgun sequencing contigs. While highly effective in plants, where mapping populations are often readily available, it must be noted that such an approach is limited by the overall recombination landscape, and the subsequent relationship between physical and genetic distance within a particular region of the genome. Recombination events in plants often occur in distal regions of the chromosomes, and peri-centromeric regions may require very large mapping populations to improve their resolution. In addition, recent studies have suggested that specific features of the genome, such as chromosomal inversion, translocation and duplication varying between the two parents used to generate the mapping population, may lead to errors and potentially confound genome assemblies.

BAC Pool Sequencing in Gene-rich Regions

A large number of genome assemblies have been generated with the help of physical maps and the use of a BAC-by-BAC sequencing approach. While laborious and costly, this approach still remains relevant as it offers multiple advantages over a whole-genome sequencing approach, especially in terms of assembling sequencing reads conserved in the context of a whole-genome assembly but mapping exclusively to a defined portion of a genome in the context of an individual clone assembly. Lonardi et al. proposed a modified version of clone sequencing to take advantage of the massive sequencing capacity offered by NGS platforms. In that study, subsets of overlapping genome-tiling BAC clones were selected and pooled according to a multidimensional grid design. Each pool then was sequenced on an Illumina HiSeq2000 instrument. The resulting paired-end reads were deconvoluted by determining, for each read the intersection between the pool it originates from and the individual BAC clones within that same pool covering the portion of the genome the read corresponds to, based on physical map information. Once deconvolution is achieved, reads can be assembled using an NGS assembler (Velvet), to recreate the sequence of the original BAC clone. Such an approach was successfully tested in barley BAC clones selected based on BAC-unigene associations, thus suggesting that BAC pool sequencing can be used in correlation with existing physical maps to complement or correct whole-genome sequencing assemblies, offering in the process the likelihood of higher quality contig sequence assemblies in gene-rich regions of complex plant genomes.

Optical Mapping

Optical mapping is a single-molecule approach that produces fingerprints using ordered restriction maps or specific nick sites. After enzymatic treatment and subsequent incorporation of fluorescent labels, the DNA molecules are stretched on a glass surface or in a nanochannel array and directly imaged to locate regions corresponding to the restriction sites or nick sites within the molecule. Distances between those sites are then inferred to produce an optical map of the DNA molecule. Two commercial platforms currently are available, namely, the Opgen Argus and the BioNano Genomics Irys systems. Using such techniques, very large DNA molecules, in the Mb range, can be interrogated for the presence and location of short recognition sites (whose sequence will vary with the enzyme being used to treat the DNA). Consensus optical maps then can be created by determining the overlap, under highly redundant conditions, between optical maps of single DNA molecules. Such consensus maps have to take into account the possibility of errors inherent to this type of technology, including star activity and false enzyme cuts, or the possibility of chimeric maps when joining, for example, optically mapped molecules containing paralogous genomic regions.

Optical maps can be used for multiple applications, including comparative genomics and structural variation detection, as well as the development of optical map-guided genome assemblies, where the optical map is aligned and compared to in silico digested contig sequences. Optical map-guided genome assemblies can assist in building high-quality genome assemblies by providing evidence of the ordering of adjacent contigs and scaffolds, or by assessing the overall sequence accuracy of contigs and suggesting potential errors in an assembly, such as inversions, translocations or chimeric contig or scaffold sequences. The addition of optical maps to a genome assembly often results in a significant increase in the scaffold N50 value. For example, Hastie et al. used the mapping of tiling BAC clones in a 2.1 Mb highly repetitive region of Aegilops tauschii (the D-genome donor of hexaploid wheat) to correct several misassemblies and improve the assembly from 75% to 95% complete. In another study, a high-resolution optical map, spanning 91% of the maize genome, was built, and used to characterize gaps within contigs, the maize genetic-physical (FPC) map and the reference pseudomolecules. Results also suggested that the placement of 12 FPC contigs on the maize genetic-physical map required re-evaluation.

Long-range Hi-C Interactions

High-throughput Chromosome capture (Hi-C) is a method that uses cross-linking of DNA-binding protein to DNA followed by restriction digestion and self-ligation of protein-bound DNA fragments, to probe genome-wide three-dimensional chromatin interactions between chromosomal regions bound to the same proteins (such as enhancer and promoter regions). There is a statistically higher probability that those regions are located on the same chromosome rather than on different chromosomes, as expected within the context of chromosomes located in distinct three-dimensional spaces within the nucleus. As a result, a vast majority of Hi-C read pairs (where each paired reads correspond to reads that may be millions of bases apart from each other on the same chromosome) can be used to determine what two contigs can be linked together on the same chromosome, based on the Hi-C paired reads they each contain.

Burton et al. evaluated the use of Hi-C datasets for long-range scaffolding of de novo whole-genome assemblies. This approach works, first, by aligning Hi-C reads to de novo assembly contig sequences and indexing each contig to their respective chromosomes, ordering contigs within each respective chromosome group by using higher Hi-C interaction densities expected between closely located contigs, and orienting ordered contigs using the location and orientation of Hi-C reads within each contig. The approach tested on existing human and mouse contig datasets generated from next-generation shotgun and mate-pair sequencing reads showed that a vast majority of the contigs could be grouped (98.2% and 98% of all sequences, in human and mouse, respectively) and ordered (94.4% and 86.7% of all grouped sequences, in human and mouse, respectively) within individual chromosomes when combined with Hi-C sequencing reads. Similar studies, where Hi-C datasets were used to complement de novo assembly generated with next-generation sequencing reads have been performed in human and mouse.

Long-range Scaffolding

Two companies, namely, 10X Genomics and Dovetail Genomics, recently presented new ways to assemble short reads delivered by the Illumina technology. The GemCode instrument from 10X Genomics is a microfluidic device used to partition very long DNA molecules (typically 50 kb or more) into oil-based droplets and to prepare Illumina-compatible libraries in combination with "gel beads", each containing a unique 14-bp indexing barcode. Once sequencing is performed, in-house software deconvolutes the barcodes and reconstructs the sequence of the original DNA subfragments as to where they originate from on the original long DNA molecule. In contrast to 10X Genomics, Dovetail Genomics approach does not necessarily require an instrument but requires larger amount of starting material for preparing samples. Dovetail's approach works essentially by in vitro making a Hi-C library from chromatin-free purified DNA, thus recreating intramolecular interactions while reducing intermolecular ones. The resulting fragments can then be selected for mate-pair sets capturing long-range intramolecular interactions for genome scaffolding. While not yet applied on plant genome assemblies, it is presumed that the strategies and technologies highlighted above could potentially assist in grouping and ordering contigs and scaffolds from gene-rich regions of diploid plant genomes.

References

- Genomics-era-for-plants-and-crop-species-advances-made-and-needed-tasks-ahead, plant-genomics: intechopen.com, Retrieved 12 January, 2020
- Naturally occurring variation in Arabidopsis: an underexploited resource for plant genetics. Trends Plant Sci. 2000;5(1):22–29
- Genomics-of-salinity-tolerance-in-plants, plant-genomics: intechopen.com, Retrieved 18 June, 2020
- Genome mapping in plant comparative genomics. Trends in Plant Science 21: 770–780
- Introduction-to-bacterial-genomics-45302: futurelearn.com, Retrieved 31 March, 2020
- Strategies-for-sequence-assembly-of-plant-genomes, plant-genomics: intechopen.com, Retrieved 27 August, 2020
- Treangen TJ, Salzberg SL (November 2011). "Repetitive DNA and next-generation sequencing: computational challenges and solutions". Nature Reviews. Genetics. 13 (1): 36–46. doi:10.1038/nrg3117. PMC 3324860. PMID 22124482
- Micrornas-sequencing-for-understanding-the-genetic-regulation-of-plant-genomes, plant-genomics: intechopen.com, Retrieved 13 April, 2020

Plant Genetics

3 CHAPTER

Plants genetics is an interdisciplinary field that studies genes, genetic variation and heredity in plants. It further includes the study of plant genetic resources, crop plant evolution and doubled haploids. This chapter has been carefully written to provide an easy understanding of plant genetics.

Plant genetics deals with heredity in plants, specifically mechanisms of hereditary transmission and variation of inherited characteristics. Plant genetics differs from animal genetics in a number of ways: somatic mutations can contribute to the germ line more easily as flowers develop at the end of branches composed of somatic cells; polyploidy is more common; and plants additionally contain chloroplastic DNA.

Plant genetics has experienced a genuine revolution during the last 15 years. This revolution started with the complete sequencing of the *Arabidopsis* and rice genomes; however, since then a large number of other plant species have had their genomes entirely decoded. Fortunately, horticultural fruit tree species have been part of this revolution, with many fruit crops being sequenced in various dicotyledonous families, such as the Rosaceae, Rutaceae and even monocotyledons, such as Musaceae and Arecaceae. The key factor for the fruit tree genomics revolution was the introduction of next generation sequencing (NGS) technologies. NGS enables the rapid generation of a massive amount of sequencing data for a fraction of the cost of the rice sequencing project, which was undertaken using earlier technologies. This means that single laboratories or small consortia now have the ability to afford to sequence the entire genome of their favourite tree crop organism at very high sequence depth within a few months. Indeed, all fruit-bearing species genomes sequenced in the last 5 years have involved utilization of some NGS data. Fortunately, the amount of NGS data generated also coincided with progress in assembly and gene annotation algorithms, as well as database and genome querying tools.

Fruit tree genome assemblies have often involved a combination of NGS technologies with old-fashioned (but accurate and consistently yielding reads over 800 bp) Sanger sequencing technology. Methods for assembling large genomes de novo using NGS were reviewed by Henson and colleagues. Briefly, the initial step of the genome assembly process involves building small contigs of overlapping NGS and Sanger sequences. However, it is impossible to build contigs spanning entire chromosomes because sequencing technologies and assembly software are not efficient in the presence of repetitive sequences. Hence, scaffolding is necessary to connect these contigs, using sequences obtained from large insert libraries such as long-insert mate paired-end libraries or bacterial artificial chromosome (BAC)-end sequencing. Nevertheless, even with the use of such large inserts, scaffolding often reaches a point at which an increase in the amount of short read sequencing does not further improve assembly quality. A metric that is widely used to measure the quality of a genome assembly is the N50 statistic. The N50 value corresponds to a median and indicates the length for which the assembly of all scaffolds of that length or longer contains at least half the sum of the lengths of the entire scaffold assembly.

Although low-throughput methods such as fluorescence in situ hybridization can help to stitch large scaffolds together, high-density genetic maps are widely used to anchor scaffolds. Sequence-based genetic markers can be retrieved from the scaffold sequences so that the scaffolds can be placed and oriented on the linkage groups, until chromosome-length scaffolds (often called pseudo-chromosomes or super-scaffolds) are developed. Following the genome assembly, the considerable task of annotating the genome remains. This includes predicting key features such as polymorphisms, GC content, repeated sequences and genes. A suite of bioinformatics tools is available for predicting protein-coding genes and repeated sequences, based on sequence homology with other sequenced genomes and alignment of RNA sequences onto the assembly contigs.

PLANT GENETIC RESOURCES

Plant genetic resources are plant genetic materials of actual or potential value. They describe the variability within plants that comes from human and natural selection over millennia. Their intrinsic value mainly concerns agricultural crops.

According to the 1983 revised International Undertaking on Plant Genetic Resources for Food and Agriculture of the Food and Agriculture Organization (FAO), plant genetic resources are defined as the entire generative and vegetative reproductive material of species with economical and/or social value, especially for the agriculture of the present and the future, with special emphasis on nutritional plants.

In the State of the World's Plant Genetic Resources for Food and Agriculture the FAO defined Plant Genetic Resources for Food and Agriculture (PGRFA) as the diversity of genetic material contained in traditional varieties and modern cultivars as well as crop wild relatives and other wild plant species that can be used now or in the future for food and agriculture.

The first use of plant genetic resources dates to more than 10,000 years ago, when farmers selected from the genetic variation they found in wild plants to develop their crops.

As human populations moved to different climates and ecosystems, taking the crops with them, the crops adapted to the new environments, developing, for example, genetic traits providing tolerance to conditions such as drought, water logging, frost and extreme heat. These traits and the plasticity inherent in having wide genetic variability - are important properties of plant genetic resources.

In recent centuries, although humans had been prolific in collecting exotic flora from all corners of the globe to fill their gardens, it wasn't until the early 20th century that the widespread and organized collection of plant genetic resources for agricultural use began in earnest. Russian geneticist Nikolai Vavilov, considered by some as the father of plant genetic resources, realized the value of genetic variability for breeding and collected thousands of seeds during his extensive travels to establish one of the first gene banks. Vavilov inspired the American Jack Harlan to collect seeds from across the globe for the United States Department of Agriculture (USDA). David Fairchild, another botanist at USDA, successfully introduced many important crops (e.g. cherries, soybeans, pistachios) into the United States.

It wasn't until 1967 that the term genetic resources was coined by Otto Frankel and Erna Bennett at the historic International Conference on Crop Plant Exploration and Conservation, organized by the FAO and the International Biological Program (IBP) "The effective utilization of genetic resources requires that they are adequately classified and evaluated" was a key message from the conference.

Conservation

Plant genetic resource conservation has become increasingly important as more plants have become threatened or rare. At the same time, an exploding world population and rapid climate change have led humans to seek new resilient and nutritious crops. Plant conservation strategies generally combine elements of conservation on farm (as part of the crop production cycle, where it continues to evolve and support farmer needs), ex situ (for example in gene banks or field collections as seed or tissue samples) or in situ (where they grow in the wild or protected areas). Most in situ conservation concerns crop wild relatives, an important source of genetic variation to crop breeding programs.

Plant genetic resources that are conserved by any of these methods are often referred to as germplasm, which is a shorthand term meaning "any genetic materials". The term originates from germ plasm, August Weismann's theory that heritable information is transmitted only by germ cells, and which has been superseded by modern insights on inheritance, including epigenetics and non-nuclear DNA.

After the Second World War, efforts to conserve plant genetic resources came mainly from breeders' organizations in the USA and Europe, which led to crop-specific collections primarily located in developed countries (e.g. IRRI, CIMMYT). In the 1960s and

1970s, more focus was put on the collection and conservation of plant genetic resources in face of genetic erosion by organizations such as the Rockefeller Foundation and the European Society of Breeding Research (EUCARPIA).

A key event in the conservation of plant genetic resources was the establishment of the International Board for Plant Genetic Resources (IBPGR) (now Bioversity International) in 1974, whose mandate was to promote and assist in the worldwide effort to collect and conserve the plant germplasm needed for future research and production. IBPGR mobilized scientists to create a global network of gene banks, thus marking the international recognition of the importance of plant genetic resources.

In 2002, the Global Crop Diversity Trust was established by Bioversity International on behalf of the CGIAR and the FAO through a Crop Diversity Endowment Fund. The goal of the Trust is to provide a secure and sustainable source of funding for the world's most important crop collections.

Policy

In response to the growing awareness of the global value of and threat to biological diversity, the United Nations drafted the 1992 Convention on Biological Diversity (CBD), the first global multilateral treaty focused on the conservation and sustainable use of biodiversity. Article 15 of the CBD specified that countries have national sovereignty over their genetic resources, but that there should be facilitated access and benefit sharing (ABS) under mutually agreed terms and with prior informed consent.

Going further to protect national sovereignty of plant genetic resources, an instrumental piece of legislation, The International Treaty on Plant Genetic Resources for Food and Agriculture (ITPGRFA), was adopted by the FAO in November 2001 and came into force in 2004.

The ITPGRFA established several mechanisms under the Multilateral System, which grants free access and equitable use of 64 of the world's most important crops (Annex 1 crops) for some uses (research, breeding and training for food and agriculture). The treaty prevents the recipients of genetic resources from claiming intellectual property rights over those resources in the form in which they received them, and ensures that access to genetic resources is consistent with international and national laws. This is facilitated by the Standard Material Transfer Agreement, a mandatory contract between providers and recipients for the exchange of germplasm under the Multilateral System. The Governing Body of the treaty, through FAO as the Third Party Beneficiary, has an interest in the agreements.

The Nagoya Protocol on Access to Genetic Resources and the Fair and Equitable Sharing of Benefits Arising from their Utilization is a supplementary agreement to the Convention on Biological Diversity that was adopted in 2010 and enforced in 2014. It provides greater legal transparency to policies governing fair and equitable sharing of benefits arising from the utilization of genetic resources.

Issues and Controversies

Due to the high value and complexity of plant genetic resources and the number of parties involved globally, some issues have arisen over their conservation and use.

Much of the material for breeding programs was collected from the Southern hemisphere and sent to gene banks in the Northern hemisphere, a concern that led to more emphasis on the national sovereignty of plant genetic resources and instigated policies that addressed the imbalance.

The increased use of plant genetic information for research, for example to find genes of interest for drought tolerance, has led to controversy on whether and to what extent the genetic data (separate from the organism) are subject to the international ABS regulations.

GENETIC DIVERSITY IN PLANTS

Genetic diversity is of fundamental importance in the continuity of a species as it provides the necessary adaptation to the prevailing biotic and abiotic environmental conditions, and enables change in the genetic composition to cope with changes in the environment. Genetic Diversity in Plants presents the magnitude of genetic variation existing in plant populations. The increasing availability of PCR-based molecular markers allows the detailed analyses and evaluation of genetic diversity in plants and also, the detection of genes influencing economically important traits. Evaluation of the genetic diversity and population structure of plant germplasm enhances breeding and management practices to develop improved varieties. Various genetic tools have been successfully used to study the genetic diversity of plant species, including morphological, cytological, biochemical, and molecular markers.

Genetic diversity of sorghum (left) and millet (right) illustrates the potential of plant breeding.

Genetic diversity within a population refers to the number of different alleles (the alternate forms of genes) of all genes and the frequency with which they appear.

Variation is high when there are many different alleles of all genes and many different combinations of those alleles.

A gene pool is the collective set of alleles found in a population. Usually gene pools refer to a particular species (individuals can interbreed freely). Plant breeders can also make use of the gene pool of closely related species that can cross and produce fertile hybrids (called interspecific hybrids).

The term "bean" is used to denote the seed of a large number of related plants: broad or fava beans, common beans, runner beans, soybeans, peas, chickpeas, vetches, and lupins. These are in fact different species.

The level of genetic variation in a population is constantly changing: different alleles of a single gene can appear and disappear from time to time within a population. This means that the gene pool of a population is always dynamic.

For some characteristics, such as maize or sorghum grain colour, there is a great level of genetic variation in nature.

In biology, a species is one of the basic units of biological classification, defined as a group of organisms capable of interbreeding and producing fertile offspring, or as a separately evolving lineage with a single gene pool.

Maize ears from CIMMYT's collection, showing a wide variety of colors and shapes. CIMMYT's germplasm bank contains about 28,000 unique samples of cultivated maize and its wild relatives.

In practice defining a species is not simple (this is known as the "species problem". Some separately evolving groups (i.e. "species") may continue to interbreed to some extent, and human intervention has resulted in extraordinary changes in the appearance of plants and animals during their domestication.

Plant breeders define species as populations of organisms that have a high level of genetic similarity.

GENETIC DIVERSITY IN CROP PLANTS

Crop plant evolution either natural or human-directed, is primarily based on existing genetic diversity in the population. Diversity can be described as the degree of differentiation between or within species. Existing intra- and inter-specific differences are at the base of all crop improvement programmes. If all the individuals within the species would have been similar, then possibly there could not have been any scope for improvement in plant performances for different traits. Since the beginning of systematic plant breeding, natural variability and divergence between crops have been extensively identified and used in improvement of crop species. However, with the progress of time, natural variability got depleted due to (i) lopsided breeding practices focusing on improvement of only few traits (like yield and its component traits), (ii) frequent use of few selected genotypes as parents in varietal development programme and (iii) introduction of few outstanding lines to many countries thereby leading to increased genetic similarity between modern crop cultivars. Reduced genetic variability and diversity among crop plant species has raised serious concern among agricultural workers. With reduced genetic diversity, further improvement in crop varieties will be an arduous task. Breaking yield barriers will become difficult and plant breeders will be unable to meet the requirements arising out of ever-increasing demand on account of exploding population. Genetic diversity becomes more important in context of climatic change and associated unforeseen events as it may serve as the reservoir of many novel traits conferring tolerance to different biotic and abiotic stresses. Genetic diversity is the underlying cause of many important agriculturally important phenomena like heterosis and transgressive segregation. Diverse lines are needed for defect correction of commercial varieties and development of novel varieties. Hence, identification of diverse lines, creation of diversity and its subsequent utilization are the major goals of any crop improvement programmes. In this context, knowledge on all aspects of genetic diversity viz., factors affecting genetic diversity, different methods of diversity analysis, their measurement and the softwares for carrying statistical analysis becomes imperative in order to utilize them prudently. Many reviews have been written focusing on vital issues like changes in genetic diversity under plant breeding, genetic vulnerability of modern crop cultivars, conservation and utilization of genetic resources assessment of genetic diversity using molecular marker and measurement of genetic diversity using statistical tools. In the present review, an attempt is being made for comprehensive compilation of overall concepts in the area of genetic diversity, which could be of immense significance for extending knowledge and meaningful research.

Concept of Diversity

Diversity is the essence of biological world. No two living things (even maternal twins) are exactly similar to each other. The difference in one or a few traits of the organism is referred to as variability. In common parlance, genetic variability and genetic diversity are considered synonym to each other which is erroneous. Genetic variability is the variation in alleles of genes or variation in DNA/RNA sequences in the gene pool of a species or population. This expresses itself in terms of alternate forms in phenotype. Genetic diversity, on the other hand, is a broad term encompassing all the variability occurring among different genotypes with respect to total genetic make-up of genotypes related to single species or between species. Genetic diversity can be measured by counting the number of different genes in a gene pool, but genetic variation can only be expected to occur and cannot be measured. Genetic variability thus, can be considered as the building blocks of genetic diversity. As recognized by Convention on Biological Diversity, there are three levels of diversity. At the highest hierarchy, lies the ecosystem diversity representing variability among different communities of species. In the next level of hierarchy, lies the species diversity representing different species within a community, also referred to as species richness. Genetic diversity is referred to the diversity present within different genotypes of same species. This is due to contrasting alleles of a gene in different individuals producing contrasting phenotypes. Swingland defined genetic diversity as the variation of heritable characteristics present in a population of the same species. The variation in heritable characters may express itself in the form of altered morphology, anatomy, physiological behaviour or biochemical features. The genomic diversity can be defined as diversity at several gene-loci within an individual. Genetic diversity has received the greatest attention among agricultural workers.

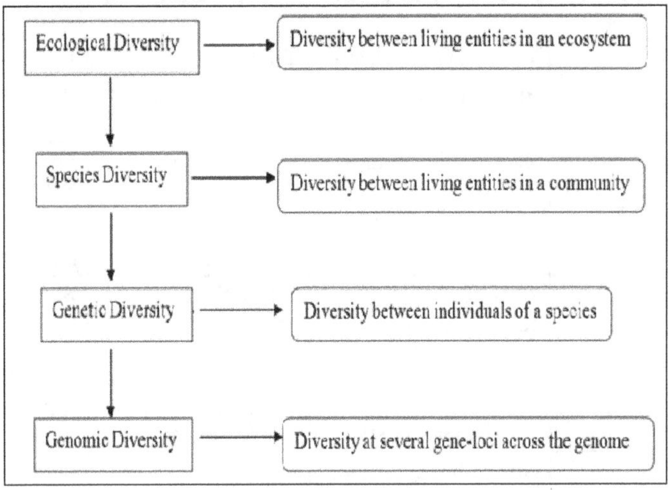

Hierarchy of diversity.

Importance of Genetic Diversity

Genetic diversity is the base for survival of plants in nature and for crop improvement. Diversity in plant genetic resources provides opportunity for plant breeders to develop new and improved cultivars with desirable characteristics, which include both farmer-preferred traits (high yield potential, large seed, etc.) and breeder-preferred traits (pest and disease resistance and photosensitivity, etc). From the very beginning of agriculture, natural genetic variability has been exploited within crop species to meet subsistence food requirement. Later the focus shifted to grow surplus food for growing populations. Presently the focus is on both yield and quality aspects of major food crops to provide balanced diet to human beings. With changing climatic scenario, breeding of climate resilient varieties is becoming more important. The existence of genetic diversity represented in the form of wild species, related species, breeding stocks, mutant lines etc. may serve as the source of desirable alleles and may assist plant breeders in breeding climate resilient varieties. The breeding of climate resilient varieties requires novel traits like tolerance towards potential new insect-pests and diseases, extreme heat, extreme cold, and towards various air- and soil-pollutants. For ever-changing breeding goals, different genes need to be reserved in cultivated and cultivable crops species in the form of germplasm resources. Presence of genetic diversity within and between crop plant species permits the breeders to select superior genotypes either to be directly used as new variety or to be used as parent in hybridization programme. Genetic diversity between two parents is essential to realize heterosis and to obtain transgressive segregants. Genetic diversity facilitates breeders to develop varieties for specific traits like quality improvement and tolerance to biotic and abiotic stresses. It also facilitates development of new lines for non-conventional uses like varieties for biofuel in sorghum, maize etc. Diversity is also important with respect to adaptability of crop plants to varied environments with special reference to changing climatic conditions. Some of the germplasm lines harbouring desirable gene in different crops are listed in table.

Table: Sources for different traits in different crop plants.

S. No.	Crop	Variety/Germplasm	Trait	Importance
1.	Okra	Pusa Sawani.	YVMV Tolerance.	Biotic stress tolerance.
2.	Maize/Sorghum	Brown mid rib lines.	Low lignin and fibre content.	Forage digestibility/ palatability.
3.	Chickpea	IPC 2004-52/PDG 84-10.	Fusarium wilt tolerance.	Biotic stress Tolerance.
4.	Cotton	Abadhita, LK 861, Kanchana, Supriya.	White fly tolerance.	Biotic stress Tolerance.
5.	Tomato	F7028.	High lycopene content.	Nutrition.
6.	Wheat	Nap Hal/UP 2672.	High protein.	Nutrition.
7.	Rice	Govindbhog/ Chakhao.	Aroma.	Palatability.

Forces affecting Genetic Diversity

Genetic diversity is primarily a function of sexual recombination. During meiosis, homologous chromosomes undergo crossing over which results in appearance of several new recombination. Different factors affect the genetic diversity in plants. Evolutionary forces like selection, mutation, migration and genetic drift act continuously and results in continuous changes in allelic frequency in a population and affects the genetic diversity. Domestication or artificial selection favours few alleles at the cost of others resulting in increased frequency of selected alleles. Consequently, domestication reduces the genetic diversity when compared to the diversity in wild. Natural selection also affects the genetic diversity considerably. Directional and stabilizing selection decreases while disruptive selection increases the genetic diversity. Mutation is also reported to increase genetic diversity. Qualitative mutation expresses itself in the form of abrupt changes in morphological/anatomical/biochemical features. Quantitative or micro-mutations have smaller and gradual effects which accumulate over time and bring about changes. Mutation may also bring about several chromosomal aberrations. Smaller sub-lethal or non-lethal aberrations bring about genetic diversity in the form of altered phenotype. Mating system of crop plants also affect genetic diversity. Inbreeding reduces while out breeding increases genetic diversity. Genetic drift can lead to loss of rare alleles thereby reduces genetic diversity. The physical distribution of individuals of a species also affects genetic diversity. Larger the physical distribution of individuals, lesser is the chances of having same genetic make-up. Some techniques like wide-hybridization, hybridization between incompatible types or introgression from previously isolated populations increase the genetic diversity as they result in generation of new phenotypes. In contrast, intra-specific hybridization reduces the genetic diversity. Gene flow within population increases the genetic diversity as new alleles are introduced.

Table: Morphological traits used as markers for different group of plants.

Group/Family	Characters
Legumes	Growth habit, Flower colour, Leaf shape, Pod and seed shape, Root Nodule traits.
Cereal crops	Stem pigmentation, Panicle length, Grain colour, Grain shape.
Fibre crops	Plant height, Basal diameter, Fibre content, Lignin and hemicellulose content.
Vegetable crops	Growth habit, Hypocotyl colour, Stem and leaf pubescence.
Forage crop	Plant height, Lignin and hemicellulose content.
Mustard	Siliqua length, Siliqua beak length, siliqua angle with main raceme, Oil content.
Medicinal crops	Herbage yield, Essential oil yield.

Methods of Diversity Analysis

Diversity analysis can be carried out using morphological, cytological, biochemical and molecular characterization. Initially, morphological markers were used for diversity analysis and are still in use. These were naturally occurring variants of a particular plant species. Later, cytological and biochemical differences occurring in the genotypes of a species started to be used in genetic diversity assessment. With the advent of genomic tools, molecular markers became the method of choice for genetic diversity assessment.

Morphological Markers

These analyses are carried out by raising germplasm lines, purelines, improved varieties etc. in a particular experimental design. This involves morphological characterization of different entries grown in the field as the morphological characteristics are the strongest determinants of the agronomic value and taxonomic classification of plants. Morphological evaluations are direct, inexpensive, easy and do not require expensive technology. However, the requirements of large tracts of land and human labourers over a period of time make it expensive. They suffer from the constraints of environmental-sensitivity and subjective characterization when compared to other methods. They are mostly dominant/recessive, have some biological effect and some morphological variants cannot survive. Different sets of morphological traits are taken into consideration for different group of crop plants.

Cytological Markers

It involves study of cytological features like chromosome size, secondary constriction in chromosomes, position of centromere, arm ratio, constitutive heterochromatic patterns, banding characteristics (G, Q, R and N banding), DNA content, total genomic chromosome length, chromosome volume etc. Different cytological features have been applied to assess genetic diversity within and between species in maize, in potato, in lentil, in radish etc. However, these have limited applications in genetic diversity analysis on account of their limited number and low resolution.

Biochemical Markers

It involves separation of proteins or their variants (isozymes) into specific banding patterns. The isozymes reflect products of different alleles and not the genes. These isozymes can be mapped onto chromosomes and can be used genetic markers for mapping other genes. This is a rapid method of assessing diversity and requires smaller amount of plant tissue as sample. However, they are limited in number, affected by environmental fluctuations and cannot be used to construct a complete genetic map.

Molecular Markers

It involves study of variation among genotypes at DNA/RNA level. Different molecular markers have different characteristics making them suitable for different purposes. They are primarily classified as hybridization-based and PCR-based. Recently, new generation of markers based on sequence or array-platforms have been developed. They can also be classified as neutral markers, genes markers and functional markers based on their activity and expression. Further, these markers may be based on variation in genomic DNA/RNA, ribosomal RNA or organelle genome sequences. Chloroplast microsatellites have been developed and used in assessment of genetic diversity at intra-specific level in wheat, barley, apple, rice, pearlmillet etc. Mitochondrial DNA in plants, in contrast, has been demonstrated to be an unsuitable tool for studying genetic diversity, being quantitatively scarce. Molecular markers are the method of choice for genetic diversity assessment on account of their hyper variability, better genomic coverage, high reproducibility, amenability to automation, being neutral and free from environmental fluctuations. Many studies on genetic diversity have been reported to use both morphological- and molecular- markers simultaneously.

Measures of Genetic Diversity

Genetic Base

Genetic base of any crop expressed in terms Coefficient of Parentage (COP) or Coefficient of Correlation. These indicate how frequently a line appears in the commercial varieties of a particular crop and is revealed by pedigree records of varieties released. COP is defined as the probability that alleles of two individuals are identical by descent. The segregating generations resulting from a cross between individuals with high COP will exhibit less variability and vice versa. The value of coefficient of parentage ranges from zero, where cultivars are completely unrelated, to one, where two cultivars have all alleles in common. The COP data matrix can be used to cluster genotypes and produce genealogically similar groups. Coefficient of parentage (COP) or coefficient of correlation (rxy) can be computed for all pairwise combinations of genotypes from pedigree information by formula given below:

$$r_{xy} = 2f_{xy} / \sqrt{(1+F_x)(1+F_y)}$$

Where, fxy is a coefficient of co-ancestry, and Fx and Fy are inbreeding coefficients of X and Y, respectively. Delannay et al., Murphy et al. and Cox et al. developed different algorithms for calculation of coefficient of parentage. Other related measure is 'Relative Genetic Contribution (RGC)' computed by partitioning the genetic constitution of a selection into theoretical percentage attributable to different ancestors. The mean genetic contribution of a given ancestor is estimated by the mean of the relative genetic contributions of this ancestor to all varieties released. The successive summation of the mean relative genetic contributions generates cumulative relative genetic contributions over times. The assumptions underlying measure of relative genetic contribution are:

- Unrelatedness of ancestors.
- Transmission of 50% of parental genes to the progeny with equal probability.

Studies on many crops revealed narrow genetic base in the released varieties of many crops in India. For example, the lines IR-8 and TN-1 (in rice), Spanish improved (in ground nut), Bragg (in soybean), T-1 and T-190 (in pigeonpea) and Pb-7 (in chickpea) appeared most frequently in commercial varieties of the respective crops released in India. Such frequent appearance of particular lines roughly gives the estimation of genetic base and consequently of genetic diversity.

Genetic Distance

Genetic distance was first defined by Nei as the difference between two entities that can be described by allelic variation. This definition was later modified to "extent of gene differences among populations that are measured using numerical values". Beumont et al. provided a more comprehensive definition of genetic distance as any quantitative measure of genetic difference at either sequence or allele frequency level calculated between genotype individuals or populations. In simple terms, genotypes with many similar genes have smaller genetic distance between them. Euclidean or straight-line measure of distance is the most commonly used statistic for estimating genetic distance between individuals (genotypes or populations) by morphological data. Mohammadi & Prasanna have described in different measures of genetic distance in detail. Euclidean distance between two genotypes can be defined mathematically as below:

$$d(a,b) = \sum_{i=0}^{n} \sqrt{[Xi - Yi]x^{2}}$$

Where, d (a,b) is the Euclidean distance between genotype a and b; Xi is the observation on i^{th} phenotypic character, and Yi is the observation on i^{th} phenotypic character. Smith et al. developed another measure of genetic diversity in inbred lines which can be expressed as below:

$$d(a,b) = \sum \sqrt{[X1_{(i)} - X2_{(i)}]^{2} / VarX_{(i)}}$$

Where, d (a,b) is the Euclidean distance between genotype a and b; X1 and X2 are the values for ith trait for inbred lines a and b and Var X(i) is the variance for ith trait over all inbred.

Genetic distances can be measured in molecular marker data where PCR amplification follows allele/locus model in following ways:

$$d(a,b) = cons\tan t \sqrt{\sum ([Xai - Xaj]^{2})}$$

Where, d (a,b) is the Euclidean distance between genotype a and b; Xai is the frequency of the allele a for individual i; Xaj is the frequency of the allele a for individual j and r is the constant based on coefficient used.

Allelic Diversity

Allelic diversity is used when genetic marker data or molecular marker data can be interpreted by locus/allele model. In such cases, data is used to generate binary matrix for further analysis. Allelic diversity can be described by (i) the percentage of polymorphic loci (p), (ii) mean number of alleles per locus (n), (iii) total gene diversity or average expected heterozygosity (H), and (iv) polymorphism information content (PIC). Percentage of polymorphic loci (p) gives an estimate of number of polymorphic loci with respect to total loci including polymorphic and monomorphic loci and can be expressed as:

$$P = \frac{N_p}{N_t} \times 1000$$

Where, Np is the number of polymorphic loci and Nt is the number of total loci (polymorphic and monomorphic).

Mean number of alleles per locus (n) is calculated by dividing total number of alleles by the number of loci and can be expressed as:

$$n = (1/k)\sum_{i=1}^{k} n_i$$

Where, k is the number of loci, and n_i is the number of alleles at i^{th} locus.

Polymorphism information content (PIC) is an indirect estimate of number of alleles per locus. This can be expressed as below:

$$PIC = 1 - \sum_{i=1}^{n} (P_i)^2$$

Where, P_i is the frequency of ith allele at any particular locus.

Estimation of Genetic Diversity using Statistical Tools

Multivariate statistics are used to assess genetic diversity among different strains/varieties/entries of a species. These techniques have a very sound theoretical basis to provide most reliable information regarding the real genetic distances between genotypes and thus can be used for assessment of genetic diversity. These techniques can be used in assessment of genetic divergence, classification of germplasm into different groups and in selection of diverse parents to develop transgressive segregants.

Metroglyph Analysis

Anderson developed a semi-graphical approach for displaying genetic diversity among a number of lines referred to as 'Metroglyph analysis'. This method represents each genotype by a circle of fixed radius (called glyph) with rays emanating from its periphery. Each variable is assigned a position on the glyph. The length of the ray represents index score of the variate. This method uses a range of variations arising from trait such that extent of trait variation is determined by the length of rays on the glyph. The performance of a genotype is adjudged by the value of the index score of that genotype. The score value determines the length of ray which may be small, medium or long.

D2 Statistics

This technique also called Mahalanobis' generalized distance was developed by Mahalanobis. This technique reduces the number of comparisons among genotypes by classifying them into different clusters. D2 values are estimated by transforming correlated variables into uncorrelated variables using pivotal condensation method. In general, the Mahalanobis distance is a measure of distance between two points in the space defined by two or more correlated variables. For example, if there are two variables that are uncorrelated, then we could plot points in a standard two-dimensional scatterplot; the Mahalanobis distances between the points would then be identical to the Euclidean distance. If there are three uncorrelated variables, we could also simply use a ruler (in a 3-D plot) to determine the distances between points. If there are more than 3 variables, we cannot represent the distances in a plot any more. In those cases, the simple Euclidean distance is not an appropriate measure, while the Mahalanobis distance will adequately account for the correlations.

Cluster Analysis

This analysis assumes discontinuities within the data. It depicts the pattern of relatedness between genotypes based on evolutionary relationships or phenotypic performance. It is used to group similar lines/germplasm in one group and differentiate other groups. It is based on methods namely (i) Unweighted paired group method using arithmetic mean (UPGMA), (ii) Unweighted paired group method using centroid (UPGMC), (iii) Weighted paired group method using arithmetic mean (WPGMA), (iv) single linkages (SLCA), (v) complete linkage (CLCA) and (vi) Median linkage (MLCA). UPGMA and UPGMC provide more accurate grouping information on breeding materials used in accordance with pedigrees and calculated results found most consistent with known heterotic groups than the other clusters.

Principal Component Analysis

Principal components analysis (PCA) can be defined as a data reduction technique applicable to quantitative type of data. PCA transforms multi-correlated variables into another set of uncorrelated variables for further study. These new set of variables are linear combinations of original variables. It is based on the development of eigen-values and mutually independent eigen-vectors (principal components) ranked in descending order of variance size. Such components give scatter plots of observations with optimal properties to study the underlying variability and correlation. Suppose x1, x2,.......,x_n be the original data in a study, then principal components may be defined as:

$$z_1 = a_{11}x_1 + a_{12}x_1 + ... + a_n x_n$$

Similarly other principal components can be defined as:

$$z_p = a_{p1}x_1 + a_{p2}x_2 + ... + a_{pn}x_n$$

This technique is not an end rather a mean for further analysis. This technique does not require any statistical model or assumption about distribution of original variate. It is worth mentioning that when original variables are uncorrelated then there is no need to carry out this analysis. This is most suitable when different variables have same unit. The difficulty of different scales can be avoided by standardizing all the variables. Standardization is done by dividing each variable by its estimated standard deviation. Recently, a spurt has been reported in the use of PCA in genetic diversity studies.

Principal Coordinate Analysis

It is another ordination method, somewhat similar to PCA, was developed by Schoenberg . The PCoA routinely finds the eigen-values and eigen-vectors of a matrix containing the distances between all data points, measured with the Gower distance or the Euclidean distance. It produces a 2 or 3 dimensional scatter plot of the samples such that the distances among the samples in the plot reflect the genetic distances among them with a minimum of distortion. This suffers from the disadvantages of (i) not providing a direct link between the components and the original variables and (ii) being complex functions of the original variables.

Canonical Analysis

Bartlett was the first to give the idea of canonical analysis. It assumes additivity in all characters and improves prediction by eliminating linear correlations between characters. Hotelling proposed the technique to describe the dependencies between two sets of variants. Seal defined it as 'a procedure of discriminating as clearly as possible between two or more multivariate normal universes with the same variance-covariance matrix'. This method has the advantage of being neutral to scale. Further, comparison of group of variables is easier when compared to that in PCA.

Factor Analysis

This technique reduces data into smaller meaningful groups based on their inter-correlations or shared variance. It is based on the assumption that correlated variables variables measure a similar factor or trait. It is used to describe the covariance relationships among many variables in terms of few underlying random quantities called factors. The main goal of factor analysis is to explain as much variance as possible in a data set by using the smallest number of factors and the smallest amount of items or variables within each factor. For interpretation of analysis, the factors with Eigen values greater than 1.0 are considered.

Correspondence Analysis

Correspondence analysis (CA) is an ordination method, somewhat similar to PCA, but for counted or discrete data. It uses Chi-square distance between the objects under study. Correspondence analysis can compare associations containing counts of taxa or counted taxa across associations. Different methods of genetic diversity analysis have been found to give similar results and hence can be used interchangeably. Chandra compared two methods and found strikingly similarity in grouping pattern of flax genotypes. On this basis, he suggested that metroglyph analysis can be used for preliminary grouping in large number of germplasms. Ariyo compared the extent of genetic diversity in okra using factor, principal component and canonical analysis and found similar results between factor and principal component analysis.

Softwares for Genetic Diversity Analysis

Many types of software have been developed for analyzing genetic diversity. Most of these softwares are based on multivariate statistics. Most of the softwares are freely available on internet and suitable for PCs. Tanavar et al. have described different programs available. Some of the softwares are briefed below:

SAS

SAS offers the package for different multivariate techniques. It involves canonical correlation, correspondence analysis, cluster analysis, factor analysis, principal component analysis etc. Principal component analysis can be performed using PROC PRINCOMP or PROC PRINQUAL. PROC CORRESP, PROC CANCORR and PROC FACTOR can be used for performing correspondence analysis, canonical correlation analysis and factorial analysis, respectively.

SPAR 3.0

IASRI, New Delhi have designed Statistical Package for Agricultural Research (SPAR). Apart from other modules, it is also capable of carrying out multivariate statistics.

Past

Paleontological Statistics software was developed by Hammer et al. It is a free, user friendly and comprehensive package. Functions found in PAST include parsimony analysis with cladogram plotting, detrended correspondence analysis, principal component analysis, principal coordinates analysis, time-series analysis, geometrical analysis etc.

NTSYSpc: Numerical Taxonomy System for Personal Computer

It is a popular program used to analyze genetic diversity from molecular marker data and has been used in different areas of science. It is based on similiarity indices and works on 0, 1 matrix of genotypic data. It is used for several applications namely cluster analysis, principal component analysis, principal coordinate analysis, etc.

GenAlEx: Genetic Analysis in Excel

It is an Excel-based and user-friendly program. It was designed for the use of SSR, SNP, AFLP, allozyme, multi locus markers and sequencing DNA data in diversiry genetics analyses. It accepts three types data viz., codominant data, dominant, and geographic data. GenALEx analysis include frequency by Locus, observed and expected heterozygosity, marker index, fixation index, Allelic Patterns, Allele list, Private alleles list, Haploid diversity by Population, Haploid diversity by Locus, Haploid disequilibrium and Pairwise Fst), Nei's Genetic Distance, Principal component analysis, Shannon index etc.

Popgene

It is another user-friendly package for the analysis of genetic diversity among and within natural populations. It enables to perform complex analysis and produce scientifically sound statistics and analyze population genetic structure using the target markers/traits. It accepts three types of data viz., codominant data, dominant and quantitative traits. The analysis include gene frequency, allele number, effective allele number, polymorphic loci, gene diversity, Shannon index, homozygosity test, F-statstics, gene flow, genetic distance (based on Nei cofficent) and dendrogram (based on UPGMA and neighbor-joining method) and neutrality.

Power Marker

It is a new program, with the first official version released in January 2004. It was designed specifically for the use of SSR/SNP data in population genetics analyses. Data can be imported from Excel or other formats, making data set-up very easy.

Importance of Genetic Diversity in Agriculture

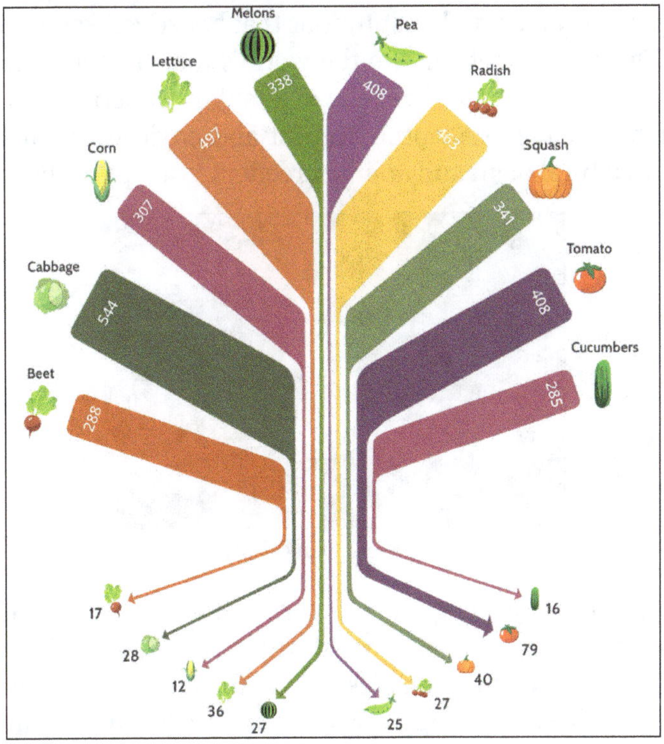

According to the International Development Research Center, the global food supply "depends on about 150 plant species. Of those, just 12 provide three-quarters of the world's food. More than half of the world's food energy comes from a limited number of varieties of three "mega-crops": rice, wheat, and maize". The global food supply is increasingly under threat from climate change, world population growth and the introduction or range expansion of disease and insects. Genetic diversity is needed to safeguard potentially vital traits that could be used to combat an unexpected future pest or adapt to the needs of the world's food supply. Plant breeders utilize genetic diversity to create improved crop varieties with traits such as yield, pest resistance and environment stress.

Left: Rice, Middle: Corn, Right: Wheat.

According to the United Nations Food and Agriculture Organization, 75% of all crop genetic diversity has been lost since the previous century, primarily due to changes in the agricultural food system which values uniformity. Of the remaining 25%, one third is expected to become extinct by 2050. Genetic diversity is the total number of genetic characteristics in the genome of a species. Curators of crop genetic diversity include seed banks, gene banks and agricultural research institutions that house collections of seeds and propagative material. These collections house the world's tools to cope with new challenges within our food system. Agricultural genetic diversity is imperative to provide a robust food security system capable of adapting to pest and environmental stressors. Genetic diversity allows agricultural plant and animal breeders to adapt to changing variables.

Gene Bank.

To adequately store genetic diversity, it is important the germplasm collections are well maintained, and backup collections are created to ensure its survival in case of a natural disaster or political unrest. Seeds and plant tissues are stored in-situ, meaning stored in a controlled environment such as cold storage or tissue culture facility. In-situ storage requires a relatively small space to accommodate a large amount of material. While propagative material from plants such as trees, shrubs and clonally propagated crops are stored ex-situ, meaning crops are preserved in an outside environment. Ex-situ storage requires a lot of space and are vulnerable to pests, natural disasters and human nature. These crops are the most at risk for extinction and most collections remain largely unstudied.

There are partnerships with private and governmental seed bank institutions to preserve crops that can be stored in-situ for long periods of time in the form of seeds at the Svalbard Global Seed Bank located north of the Arctic Circle. The Svalbard Global Seed Bank acts as a safety deposit box which allows institutions to safety deposit and withdraw their backup seed collections in case their main collection is lost or severely damaged. There are some regional and national partnerships regarding ex-situ crops in the United States and abroad, but international partnerships are difficult to conduct due to plant quarantine restrictions and the possible importation of pests. Until new efficient methods are devised to limit the transfer of pests, there is a need for increased funding and effort to create more backup collection sites spread throughout the world. It is important that the remaining uncollected crop genetic diversity from wild crop relatives, unique plant populations with low representation in gene banks and minor crops located in remote regions of the world, are collected and preserved before they go extinct. Wild crop relatives are the ancestors of our domesticated crops. They may contain important genetic traits, which may have been lost during the domestication process. Traits such as drought or pest resistance will be crucial in adapting to climate change. Many wild crop species are at risk for extinction due to habitat destruction or the encroachment of urban sprawl. Unique plant populations that are currently underrepresented in gene banks need to be collected due to current material limited to a few individuals, which does not adequately represent all the genetic diversity within that population. These populations are located in small pockets, which are extremely vulnerable to extinction due to pests or human activity.

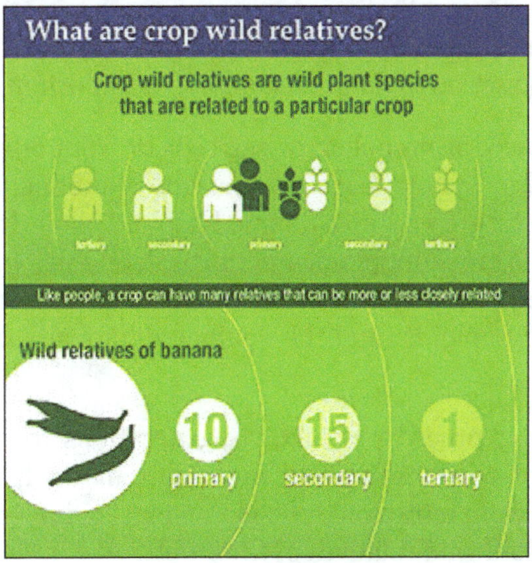

The Plant Germplasm System currently houses almost 580,000 accessions of crop material. In the last 8 years it has gained almost 60,000 more accessions. An accession is an individual cultivar or variety of a crop. During the same period, the budget has remained relatively stagnant. With such a large increase in the amount of accessions being curated, resources devoted to important programs such as plant expeditions, infrastructure, regeneration, and development of improved conservation practices are strained. Genetic diversity is an important resource that needs to be properly collected and conserved for future generations.

Potato Diveristy.

Genetic diversity will play a crucial role in the development of crops adapted to climate change and the production of food for the growing world population. To ensure these resources are available to plant breeders in the future, more public funding in seed banks and agricultural institutions is needed. The public needs to be informed of the importance of our genetic resources for the sake of improving the global food system.

DOUBLED HAPLOID

A doubled haploid (DH) is a genotype formed when haploid cells undergo chromosome doubling. Artificial production of doubled haploids is important in plant breeding.

Haploid cells are produced from pollen or egg cells or from other cells of the gametophyte, then by induced or spontaneous chromosome doubling, a doubled haploid cell is produced, which can be grown into a doubled haploid plant. If the original plant was diploid, the haploid cells are monoploid, and the term doubled monoploid may be used for the doubled haploids. Haploid organisms derived from tetraploids or hexaploids are sometimes called dihaploids (and the doubled dihaploids are, respectively, tetraploid or hexaploid).

Conventional inbreeding procedures take six generations to achieve approximately complete homozygosity, whereas doubled haploidy achieves it in one generation. Dihaploid plants derived from tetraploid crop plants may be important for breeding programs that involve diploid wild relatives of the crops.

Production of Doubled Haploids

Doubled haploids can be produced in vivo or in vitro. Haploid embryos are produced in vivo by parthenogenesis, pseudogamy, or chromosome elimination after wide crossing. The haploid embryo is rescued, cultured, and chromosome-doubling produces doubled haploids. The in vitro methods include gynogenesis (ovary and flower culture) and androgenesis (anther and microspore culture). Androgenesis is the preferred method. Another method of producing the haploids is wide crossing. In barley, haploids can be produced by wide crossing with the related species Hordeum bulbosum; fertilization is affected, but during the early stages of seed development the H. bulbosum chromosomes are eliminated leaving a haploid embryo. In tobacco (Nicotiana tabacum), wide crossing with Nicotiana africana is widely used. When N. africana is used to pollinate N. tabacum, 0.25 to 1.42 percent of the progeny survive and can readily be identified as either F1 hybrids or maternal haploids. Although these percentages appear small, the vast yield of tiny seeds and the early death of most seedlings provide significant numbers of viable hybrids and haploids in relatively small soil containers. This method of interspecific pollination serves as a practical way of producing seed-derived haploids of N. tabacum, either as an alternative method or complementary method to anther culture.

Genetics of DH Population

In DH method only two types of genotypes occur for a pair of alleles, A and a, with the frequency of ½ AA and ½ aa, while in diploid method three genotypes occur with the frequency of ¼ AA, ½ Aa, ¼ aa. Thus, if AA is desirable genotype, the probability of obtaining this genotype is higher in haploid method than in diploid method. If n loci are segregating, the probability of getting the desirable genotype is (1/2)n by the haploid method and (1/4)n by the diploid method. Hence the efficiency of the haploid method is high when the number of genes concerned is large.

Studies were conducted comparing DH method and other conventional breeding methods and it was concluded that adoption of doubled haploidy does not lead to any bias of genotypes in populations, and random DHs were even found to be compatible to selected line produced by conventional pedigree method.

Applications of DHs Plant Breeding

Mapping Quantitative Trait Loci

Most of the economic traits are controlled by genes with small but cumulative effects. Although the potential of DH populations in quantitative genetics has been understood for some time, it was the advent of molecular marker maps that provided the impetus for their use in identifying loci controlling quantitative traits. As the quantitative trait loci (QTL) effects are small and highly influenced by environmental factors, accurate phenotyping with replicated trials is needed. This is possible with doubled haploidy organisms because of their true breeding nature and because they can conveniently be produced in large numbers. Using DH populations, 130 quantitative traits have been mapped in nine crop species. In total, 56 DH populations were used for QTL detection.

Backcross Breeding

In backcross conversion, genes are introgressed from a donor cultivar or related species into a recipient elite line through repeated backcrossing. A problem in this procedure is being able to identify the lines carrying the trait of interest at each generation. The problem is particularly acute if the trait of interest is recessive, as it will be present only in a heterozygous condition after each backcross. The development of molecular markers provides an easier method of selection based on the genotype (marker) rather than the phenotype. Combined with doubled haploidy it becomes more effective. In marker assisted backcross conversion, a recipient parent is crossed with a donor line and the hybrid (F1) backcrossed to the recipient. The resulting generation (BC1) is backcrossed and the process repeated until the desired genotypes are produced. The combination of doubled haploidy and molecular marker provides the short cut. In the back cross generation one itself a genotype with the character of interest can be selected and converted into homozygous doubled haploid genotype. Chen et al. used marker assisted backcross conversion with doubled haploidy of BC1 individuals to select stripe rust resistant lines in barley.

Bulked Segregant Analysis (BSA)

In bulked segregant analysis, a population is screened for a trait of interest and the genotypes at the two extreme ends form two bulks. Then the two bulks are tested for the presence or absence of molecular markers. Since the bulks are supposed to contrast in the alleles that contribute positive and negative effects, any marker polymorphism between the two bulks indicates the linkage between the marker and trait of interest. BSA is dependent on accurate phenotyping and the DH population has particular advantage in that they are true breeding and can be tested repeatedly. DH populations are commonly used in bulked segregant analysis, which is a popular method in marker assisted breeding. This method has been applied mostly to rapeseed and barley.

Genetic Maps

Genetic maps are very important to understand the structure and organization of genomes from which evolution patterns and syntenic relationships between species can be deduced. Genetic maps also provide a framework for the mapping of genes of interest and estimating the magnitude of their effects and aid our understanding of genotype/phenotype associations. DH populations have become standard resources in genetic mapping for species in which DHs are readily available. Doubled haploid populations are ideal for genetic mapping. It is possible to produce a genetic map within two years of the initial cross regardless of the species. Map construction is relatively easy using a DH population derived from a hybrid of two homozygous parents as the expected segregation ratio is simple, i.e. 1:1. DH populations have now been used to produce genetic maps of barley, rapeseed, rice, wheat, and pepper. DH populations played a major role in facilitating the generation of the molecular marker maps in eight crop species.

Genetic Studies

Genetic ratios and mutation rates can be read directly from haploid populations. A small doubled haploid (DH) population was used to demonstrate that a dwarfing gene in barley is located chromosome 5H. In another study the segregation of a range of markers has been analyzed in barley.

Genomics

Although QTL analysis has generated a vast amount of information on gene locations and the magnitude of effects on many traits, the identification of the genes involved has remained elusive. This is due to poor resolution of QTL analysis. The solution for this problem would be production of recombinant chromosome substitution line, or stepped aligned recombinant inbred lines. Here, backcrossing is carried out until a desired level of recombination has occurred and genetic markers are used to detect desired recombinant chromosome substitution lines in the target region, which can be fixed by doubled haploidy. In rice, molecular markers have been found to be linked with major genes and QTLs for resistance to rice blast, bacterial blight, and sheath blight in a map produced from DH population.

Elite Crossing

Traditional breeding methods are slow and take 10–15 years for cultivar development. Another disadvantage is inefficiency of selection in early generations because of heterozygosity. These two disadvantages can be over come by DHs, and more elite crosses can be evaluated and selected within less time.

Cultivar Development

Uniformity is a general requirement of cultivated line in most species, which can be easily obtained through DH production. There are various ways in which DHs can be used in cultivar production. The DH lines themselves can be released as cultivars, they may be used as parents in hybrid cultivar production or more indirectly in the creation of breeders lines and in germplasm conservation. Barley has over 100 direct DH cultivars. According to published information there are currently around 300 DH derived cultivars in 12 species worldwide.

The relevance of DHs to plant breeding has increased markedly in recent years owing to the development of protocols for 25 species. Doubled haploidy already plays an important role in hybrid cultivar production of vegetables, and the potential for ornamental production is being vigorously examined. DHs are also being developed in the medicinal herb Valeriana officinalis to select lines with high pharmacological activity. Another interesting development is that fertile homozygous DH lines can be produced in species that have self-incompatibility systems.

Advantages of DHs

The ability to produce homozygous lines after a single round recombination saves a lot of time for the plant breeders. Studies conclude that random DH's are comparable to the selected lines in pedigree inbreeding. The other advantages include development of large number of homozygous lines, efficient genetic analysis and development of markers for useful traits in much less time. More specific benefits include the possibility of seed propagation as an alternative to vegetative multiplication in ornamentals, and in species such as trees in which long life cycles and inbreeding depression preclude traditional breeding methods, doubled haploidy provides new alternatives.

Disadvantages of DHs

The main disadvantage with the DH population is that selection cannot be imposed on the population. But in conventional breeding selection can be practised for several generations: thereby desirable characters can be improved in the population.

In haploids produced from anther culture, it is observed that some plants are aneuploids and some are mixed haploid-diploid types. Another disadvantage associated with the double haploidy is the cost involved in establishing tissue culture and growth facilities. The over-usage of doubled haploidy may reduce genetic variation in breeding germplasm. Hence one has to take several factors into consideration before deploying doubled haploidy in breeding programmes.

HEREDITY IN PLANTS

One of the key things we see in the living world all around us is the principle of heredity: the rule that like begets like. A seed from an apple gives rise to another apple tree, not an orange tree or a beanstalk. Children invariably resemble their parents, both in looks and in personality. How does this come about? How does the tiny seed 'know' that it has to grow into an apple tree and not anything else? Clearly, there must be some medium for the transmission of information from plant to seed, or parent to child. The concept of a gene as the basic unit of this medium was first proposed by an Austrian priest who enjoyed experimenting with pea plants in his backyard.

In his most famous experiments, done between 1856–63, Mendel took two different varieties of the plant, a tall one and a short one, and produced offspring by hybridising them (i.e., using pollen from one to fertilise the other). According to the blending inheritance theory popular at that time, the progeny should have been of medium height, intermediate between the characteristics of the parents. However, Mendel found that the progeny in the first generation, known as the F1 (first filial) generation, were always tall. When these hybrids were further interbred, to produce second generation or F2 progeny, the short variety reappeared, but only 1 in 4 (or 25%) of the F2 plants were short and the rest were tall.

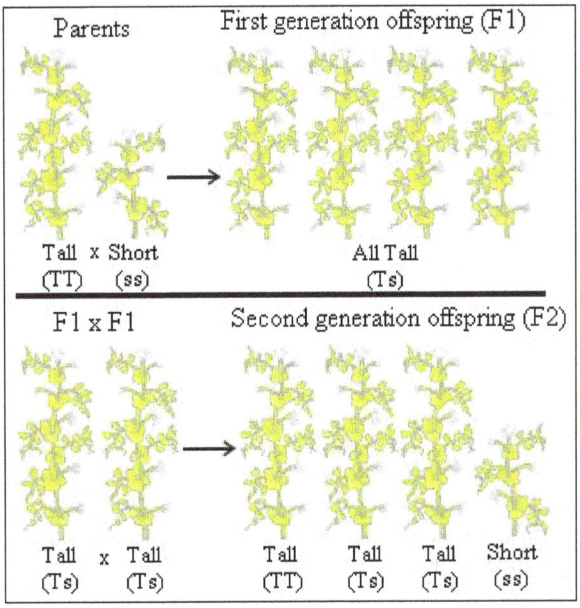

Based on his experiments, Mendel proposed the idea that biological traits were inherited discretely for instance, the height of the pea plant would take on one of only two values, 'tall' or 'short', rather than being able to vary continuously over a range of values. The term 'gene' was coined for a unit of heredity that determined a single biological trait; at that time people had no idea what genes were actually made of or where they were located, but we now know that they are composed of a molecule called deoxy-ribonucleic acid (DNA), and are found inside the cell nucleus. According to Mendel's theory, a gene could have several types, known as alleles: the gene for height in pea plants would have two types, tall and short. Also, each individual would have two copies of a given gene, since one would be inherited (at random) from each parent. What if the two copies were of different types? Mendel's results suggested that in such cases only one type would actually be expressed biologically meaning the effects of only one would be observed in the individual's physical characteristics, whilst the other would remain hidden. The former type is known as the dominant allele, and the latter as the recessive allele. In the above diagram, the dominant type for the height gene is 'tall', denoted by 'T', whilst the recessive type is 'short', denoted by 's'.

We use the terms genotype and phenotype to distinguish between what genes an individual has and what its physical characteristics are. Genotype simply means the types of the two copies for any gene, for example, all the F1 plants have genotype 'Ts'. Phenotype means the actual properties seen for that trait: the F1 plants above all have the phenotype 'Tall'.

Inferrring Allele Dominance

In the example of Mendel's experiment above, he could infer that the tall type is dominant because he knew that the original plants used by him were purebred, meaning that both their gene copies were of the same type. However, in general a single inheritance event may not provide enough information to infer which of two types of a given trait is dominant. For instance, if we find that a tall plant hybridised with a short plant to produce a tall offspring, both of these scenarios are possible:

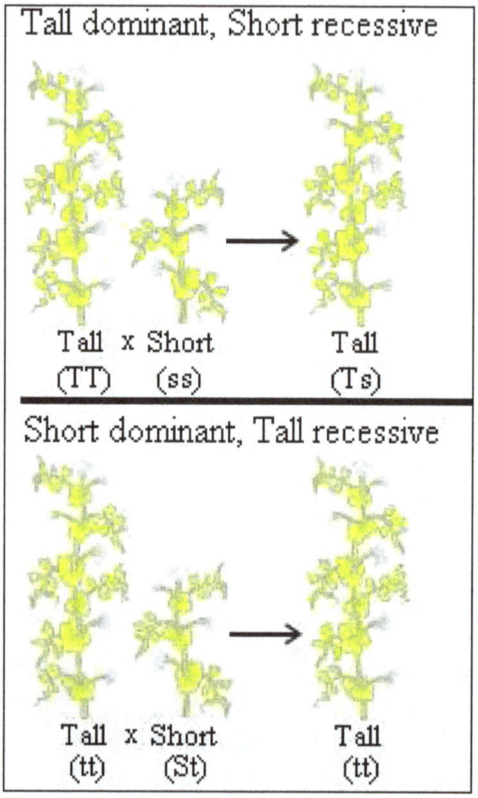

On the other hand, if two short plants produce a tall offspring, then we can be sure the short type must be dominant. Consider the alternative: supposing the tall type is dominant. Then both parents must have the genotype 'ss', as even a single copy of 'T' would have made them tall. This means the child must inherit 's' from both parents, and therefore also have genotype 'ss' and phenotype 'Short'. So if we observe the child to be tall, this possibility is ruled out and short must be dominant. Similarly, two tall plants producing a short child would imply that tall is dominant.

Mendel studied the inheritance of seven different features in peas, including height, flower color, seed color, and seed shape. To do so, he first established pea lines with two different forms of a feature, such as tall vs. short height. He grew these lines for generations until they were pure-breeding (always produced offspring identical to the parent), then bred them to each other and observed how the traits were inherited.

In addition to recording how the plants in each generation looked, Mendel counted the exact number of plants that showed each trait. Strikingly, he found very similar patterns of inheritance for all seven features he studied:

- One form of a feature, such as tall, always concealed the other form, such as short, in the first generation after the cross. Mendel called the visible form the dominant trait and the hidden form the recessive trait.
- In the second generation, after plants were allowed to self-fertilize (pollinate themselves), the hidden form of the trait reappeared in a minority of the plants. Specifically, there were always about 3333 plants that showed the dominant trait (e.g., tall) for every 1111 plant that showed the recessive trait (e.g., short), making a 3:13:13:13, colon, 1 ratio.
- Mendel also found that the features were inherited independently: one feature, such as plant height, did not influence inheritance of other features, such as flower color or seed shape.

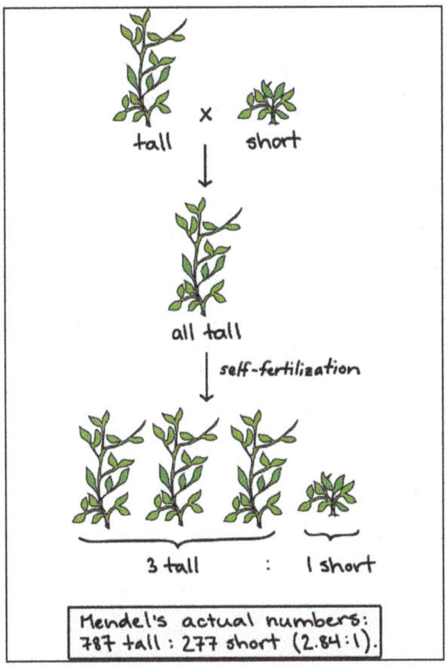

Representation of results from one of Mendel's experiments. When a tall and short plant are crossed, all of the offspring are tall. If the offspring self-fertilize, they produce tall and short plants in a ratio of 3:1 in the next generation. Mendel's actual counts were 787 tall: 277 short plants in this generation.

In 1865, Mendel presented the results of his experiments with nearly 30,000 pea plants to the local Natural History Society. Based on the patterns he observed, the counting data he collected, and a mathematical analysis of his results, Mendel proposed a model of inheritance in which:

- Characteristics such as flower color, plant height, and seed shape were controlled by pairs of heritable factors that came in different versions.

- One version of a factor (the dominant form) could mask the presence of another version (the recessive form).
- The two paired factors separated during gamete production, such that each gamete (sperm or egg) randomly received just one factor.
- The factors controlling different characteristics were inherited independently of one another.

In 1866, Mendel published his observations and his model of inheritance, under the title Experiments in Plant Hybridization3,4^{3,4}3,4start superscript, 3, comma, 4, end superscript, in the Proceedings of the Natural History Society of Brünn.

Scientific Legacy

Mendel's work went largely unnoticed by the scientific community during his lifetime.

In part, Mendel's contemporaries failed to recognize the importance of his work because his findings went against prevailing (popular) ideas about inheritance. In addition, although we now see Mendel's mathematical approach to biology as innovative and pioneering, it was new, unfamiliar, and perhaps confusing or unintuitive to other biologists of the time.

In the mid-1800s, when Mendel was doing his experiments, most biologists subscribed to the idea of blending inheritance. Blending inheritance wasn't a formal, scientific hypothesis, but rather, a general model in which inheritance involved the permanent blending of parents' characteristics in their offspring (producing offspring with an intermediate form of a characteristic). But the blending model could not explain why Mendel crossed a tall and a short pea plant and got only tall plants, or why self-fertilization of one of those tall plants would produce a 3:13:13:13, colon, 1 ratio of tall to short plants in the next generation. Instead, if the blending model were correct, a tall plant crossed with a short plant should produce a medium plant, which would go on to produce more medium plants.

Image comparing the predictions of the blending model with Mendel's actual results for a cross between a tall pea plant and a short pea plant.

The blending model predicts that all the offspring from the cross should be of medium height, and that if those offspring self-fertilize, all the plants in the next generation will also be of medium height.

Mendel instead observed that all the offspring of the cross were tall, and that when they self-fertilized, they produced tall and short plants in a ratio of 3:1.

As it turns out, both pea plant height and human height (along with many other characteristics in a wide range of organisms) are controlled by pairs of heritable factors that come in distinctive versions, just as Mendel proposed. In humans, however, there are many different factors (genes) that contribute fractionally to height and vary among individuals. This makes it difficult to see the contribution of any one factor and produces inheritance patterns that can resemble blending. In Mendel's experiments, in contrast, there was just one factor that differed between the tall and short pea plants, allowing Mendel to clearly see the underlying pattern of inheritance.

It was not until around 1900 that his work was rediscovered, reproduced, and revitalized. Its rediscoverers were biologists on the brink of discovering the chromosomal basis of heredity that is, about to realize that Mendel's "heritable factors" were carried on chromosomes.

Mendel's Model System

Mendel carried out his key experiments using the garden pea, Pisum sativum, as a model system. Pea plants make a convenient system for studies of inheritance, and they are still studied by some geneticists today.

Useful features of peas include their rapid life cycle and the production of lots and lots of seeds. Pea plants also typically self-fertilize, meaning that the same plant makes both the sperm and the egg that come together in fertilization. Mendel took advantage of this property to produce true-breeding pea lines: he self-fertilized and selected peas for many generations until he got lines that consistently made offspring identical to the parent (e.g., always short).

Pea plants are also easy to cross, or mate in a controlled way. This is done by transferring pollen from the anthers (male parts) of a pea plant of one variety to the carpel (female part) of a mature pea plant of a different variety. To prevent the receiving plant from self-fertilizing, Mendel painstakingly removed all of the immature anthers from the plant's flowers before the cross.

First, a flower on the female parent is emasculated, meaning that the male parts (anthers) are removed with forceps or scissors. Then, pollen is collected from a flower on the male parent plant using a paintbrush. The pollen is dabbed onto the female part (carpel) of the female parent flower that was previously emasculated.

Because peas were so easy to work with and prolific in seed production, Mendel could perform many crosses and examine many individual plants, making sure that his results were consistent (not just a fluke) and accurate (based on many data points).

Mendel's Experimental Setup

Once Mendel had established true-breeding pea lines with different traits for one or more features of interest (such as tall vs. short height), he began to investigate how the traits were inherited by carrying out a series of crosses.

First, he crossed one true-breeding parent to another. The plants used in this initial cross are called the P generation, or parental generation.

Mendel collected the seeds from the P generation cross and grew them up. These offspring were called the F_1 generation, short for first filial generation.

Once Mendel examined the F_1 plants and recorded their traits, he let them self-fertilize naturally, producing lots of seeds. He then collected and grew the seeds from the F_1 plants to produce an F_2 generation, or second filial generation. Again, he carefully examined the plants and recorded their traits.

Mendel's experiments extended beyond the F_2 generation to F_3, F_4 and later generations, but his model of inheritance was based mostly on the first three generations (P, F_1 and F_2).

Mendel didn't just record what his plants looked like in each generation (e.g., tall vs. short). Instead, he counted exactly how many plants with each trait were present. This may sound tedious, but by recording numbers and thinking mathematically, Mendel made discoveries that eluded famous scientists of his time (such as Charles Darwin, who carried out similar experiments but didn't grasp the significance of his results).

References

- Plant-genetics, agricultural-and-biological-sciences: sciencedirect.com, Retrieved 16 June, 2020
- Genetic-diversity, bioscience-in-brief-plantbreeding: b4fa.org, Retrieved 25 August, 2020
- Deplazes-Zemp, Anna (1 June 2018). "'Genetic resources', an analysis of a multifaceted concept" (PDF). Biological Conservation. 222: 86–94. doi:10.1016/j.biocon.2018.03.031
- Assessment-of-Genetic-Diversity-in-Crop-Plants-An-Overview-317845030: researchgate.net, Retrieved 28 January, 2020
- Importance-of-genetic-diversity-in-agriculture, thenextnorm: medium.com, Retrieved 05 May, 2020
- Genetics: web.iitd.ac.in, Retrieved 10 March, 2020
- Mendel-and-his-peas, science-high-school-biology-hs-classical-genetics-hs-introduction-to-heredity: khanacademy.org, Retrieved 13 February, 2020

Plant Reproductive Systems

CHAPTER 4

Plant reproductive system is defined as the system by which plants reproduce. It is branched into two types – sexual and asexual reproduction. In sexual reproduction the offspring is produced by the fusion of gametes. In asexual reproduction the offspring is produced without the fusion of gametes. The topics elaborated in this chapter will help in gaining a better perspective of the subject of plant reproductive system.

Plant reproductive system is any of the systems, sexual or asexual, by which plants reproduce. In plants, the end result of reproduction is the continuation of a given species, and the ability to reproduce is, therefore, rather conservative, or given to only moderate change, during evolution. Changes have occurred, however, and the pattern is demonstrable through a survey of plant groups.

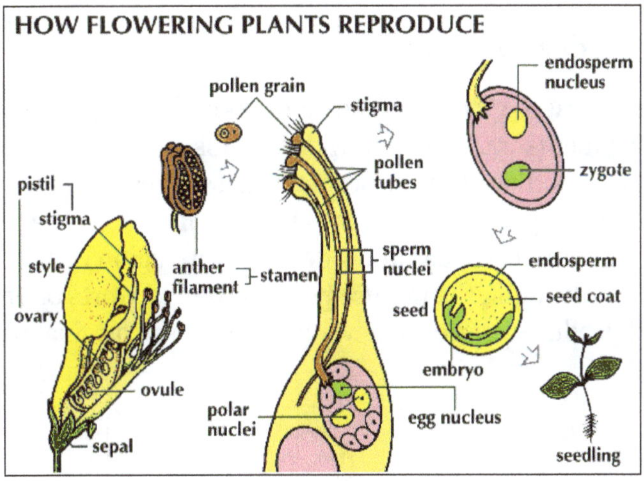

Reproduction in flowering plants begins with pollination, the transfer of pollen from anther to stigma on the same flower or to the stigma of another flower on the same plant (self-pollination) or from the anther on one plant to the stigma of another plant (cross-pollination). Once the pollen grain lodges on the stigma, a pollen tube grows from the pollen grain to an ovule. Two sperm nuclei then pass through the pollen tube. One of them unites with the egg nucleus and produces a zygote. The other sperm nucleus unites with two polar nuclei to produce an endosperm nucleus. The fertilized ovule develops into a seed.

Reproduction in plants is either asexual or sexual. Asexual reproduction in plants involves a variety of widely disparate methods for producing new plants identical in every respect to the parent. Sexual reproduction, on the other hand, depends on a complex series of basic cellular events, involving chromosomes and their genes that take place within an elaborate sexual apparatus evolved precisely for the development of new plants in some respects different from the two parents that played a role in their production.

In order to describe the modification of reproductive systems, plant groups must be identified. One convenient classification of organisms sets plants apart from other forms such as bacteria, algae, fungi, and protozoans. Under such an arrangement, the plants, as separated, comprise two great divisions (or phyla) the Bryophyta (mosses and liverworts) and the Tracheophyta (vascular plants). The vascular plants include four subdivisions: the three entirely seedless groups are the Psilopsida, Lycopsida, and Sphenopsida; the fourth group, the Pteropsida, consists of the ferns (seedless) and the seed plants (gymnosperms and angiosperms).

Asexual reproduction involves no union of cells or nuclei of cells and, therefore, no mingling of genetic traits, since the nucleus contains the genetic material (chromosomes) of the cell. They fall into two basic types: systems that utilize almost any fragment or part of a plant body and systems that depend upon specialized structures that have evolved as reproductive agents.

Reproduction by Fragments

In many plant groups, fragmentation of the plant body, followed by regeneration and development of the fragments into whole new organisms, serves as a reproductive system. Fragments of the plant bodies of liverworts and mosses regenerate to form new plants. In nature and in laboratory and greenhouse cultures, liverworts fragment as a result of growth; the growing fragments separate by decay at the region of attachment to the parent. During prolonged drought, the mature portions of liverworts often die, but their tips resume growth and produce a series of new plants from the original parent plant.

Archegonium: The female reproductive organ, or archegonium, emerging from a thalloid liverwort.

In mosses, small fragments of the stems and leaves (even single cells of the latter) can, with sufficient moisture and under proper conditions, regenerate and ultimately develop into new plants.

It is common horticultural practice to propagate desirable varieties of garden plants by means of plant fragments, or cuttings. These may be severed leaves or portions of roots or stems, which are stimulated to develop roots and produce leafy shoots. Naturally fallen branches of willows (Salix) and poplars (Populus) root under suitable conditions in nature and eventually develop into trees. Other horticultural practices that exemplify asexual reproduction include budding (the removal of buds of one plant and their implantation on another) and grafting (the implantation of small branches of one individual on another).

Reproduction by Special Asexual Structures

Throughout the plant kingdom, specially differentiated or modified cells, groups of cells, or organs have, during the course of evolution, come to function as organs of asexual reproduction. These structures are asexual in that the individual reproductive agent develops into a new individual without the union of sex cells (gametes). A number of examples of special asexual agents of reproduction from several plant groups are in this section.

Airborne spores characterize most nonflowering land plants, such as mosses, liverworts, and ferns. Although the spores arise as products of meiosis, a cellular event in which the number of chromosomes in the nucleus is halved, such spores are asexual in the sense that they may grow directly into new individuals, without prior sexual union.

Among liverworts, mosses, lycopods, ferns, and seed plants, few-to many-celled specially organized buds, or gemmae, also serve as agents of asexual reproduction.

The vegetative, or somatic, organs of plants may, in their entirety, be modified to serve as organs of reproduction. In this category belong such flowering-plant structures as stolons, rhizomes, tubers, corms, and bulbs, as well as the tubers of liverworts, ferns, and horsetails, the dormant buds of certain moss stages, and the leaves of many succulents. Stolons are elongated runners, or horizontal stems, such as those of the strawberry, which root and form new plantlets when they make proper contact with a moist soil surface. Rhizomes, as seen in iris, are fleshy, elongated, horizontal stems that grow within or upon the soil. The branching of rhizomes results in multiplication of the plant. The enlarged fleshy tips of subterranean rhizomes or stolons are known as tubers, examples of which are potatoes. Tubers are fleshy storage stems, the buds ("eyes") of which, under proper conditions, can develop into new individuals. Erect, vertical, fleshy, subterranean stems, which are known as corms, are exemplified by crocuses and gladioli. These organs tide the plants over periods of dormancy and may develop secondary cormlets, which give rise to new plantlets. Unlike the corm, only a small portion of the bulb, as in lilies and the onion, represents stem tissue. The latter is surrounded by the fleshy food-storage bases of earlier-formed leaves. After a period of dormancy, bulbs develop into new individuals. Large bulbs produce secondary bulbs through development of buds, resulting in an increase in number of individuals.

The Cellular Basis

Sexual reproduction at the cellular level generally involves the following phenomena: the union of sex cells and their nuclei, with concomitant association of their chromosomes, which contain the genes, and the nuclear division called meiosis. The sex cells are called gametes, and the product of their union is a zygote. All gametes are normally haploid (having a single set of chromosomes) and all zygotes, diploid (having a double set of chromosomes, one set from each parent). Gametes may be motile, by means of whiplike hairs (flagella) or of flowing cytoplasm (amoeboid motion). In their union, gametes may be morphologically indistinguishable (i.e., isogamous) or they may be distinguishable only on the criterion of size (i.e., heterogamous). The larger gamete, or egg, is nonmotile; the smaller gamete, or sperm, is motile. The last type of gametic difference, egg and sperm, is often designated as oogamy. In oogamous reproduction, the union of sperm and egg is called fertilization. Isogamy, heterogamy, and oogamy are often considered to represent an increasingly specialized evolutionary series.

In the plants included in this article bryophytes (mosses and liverworts) and tracheophytes (vascular plants) sexual reproduction is of the oogamous type, or a modification thereof, in which the sex cells, or gametes, are of two types, a larger nonmotile egg and a smaller motile sperm. These gametes are often produced in special containers called gametangia, which are multicellular. In cases in which special gametangia are lacking, every cell produces a gamete. In oogamy, the male gametangia are called antheridia and the female oogonia or archegonia. A female gametangium with a sterile cellular jacket is called an archegonium, although, like an oogonium, it produces eggs. In most of the plants dealt with in this article, the eggs are produced in archegonia and the sperms in antheridia with surface layers of sterile cells.

The Plant Basis

Individual plants may be either bisexual (hermaphroditic), in which male and female gametes are produced by the same organism, or unisexual, producing either male or female gametes but not both. A bisexual individual, however, is not necessarily capable of fertilizing its own eggs. In certain ferns, for example, male gametes of one individual are not compatible with the female gametes of the same individual, so cross-fertilization (with another individual of the species) is obligatory. This situation, of course, is similar in adaptive significance to cross-pollination (which leads to cross-fertilization) among seed plants.

Among the liverworts, mosses, and vascular plants, the life cycle involves two different phases, often called generations, although only one plant generation is, in fact, involved in one complete cycle. This type of life cycle is often said to illustrate the "alternation of generations," in which a haploid individual (i.e., with one set of chromosomes), or tissue, called a gametophyte, at maturity produces gametes that unite in pairs to form diploid (i.e., containing two sets of chromosomes) zygotes. The latter develop directly into individuals, or tissues, called sporophytes, in which the nuclei of certain fertile cells, called spore mother cells, or sporocytes, give rise to haploid spores (sometimes called meiospores). These spores are lightweight and are borne by air currents; they germinate to form the haploid, sexual, gamete-producing phase, usually designated the gametophyte.

There are several variations in the above-described life cycle. The haploid gametophyte and sporophyte may be free-living, independent plants (e.g., certain algae and yeasts), in which case the life cycle is diplobiontic, or the sporophyte may be physically attached to the gametophyte, as it is in liverworts and mosses. By contrast, the gametophytic phases develop as parasites on the sporophytes of the seed plants, as in certain algae. In further variation, the alternating phases may be similar morphologically except for the type of reproductive cells (gametes or spores) they produce (isomorphic life cycle), or they may be strikingly dissimilar, as in some algae, mosses, ferns, and seed plants (heteromorphic life cycle). Only heteromorphic life cycles occur in liverworts, mosses, vascular plants, and certain fungi.

The differences between the gametophyte and sporophyte are often great, especially those of the diplobiontic types, so the alternates seem to be two different, unrelated individuals rather than different manifestations of the same organism.

Bryophyte Reproductive Systems

Liverworts and Hornworts

The plant bodies of liverworts and hornworts represent the gametophytic (sexual) phase of the life cycle, which is dominant in these plants. In the liverworts, the sporophyte is borne upon or within the gametophyte but is transitory. Liverwort and hornwort plants, depending on the species, may be bisexual or unisexual, and the sex organs may be distributed on the surface (Riccia, Ricciocarpus, Sphaerocarpos, Pellia) or localized in groups and borne on special branches (antheriodiophores and archegoniophores), as in Marchantia; the sperm cells are biflagellate.

Hornwort (Dendroceros).

Release of the mature sperm and the process of fertilization require moisture in the form of heavy dew or raindrops. In all but a few genera (Riccia, Ricciocarpus), the developing sporophytes are actively photosynthetic i.e., capable of utilizing light energy to form organic substances. They are, however, dependent on gametophytic tissues for water (and the inorganic salts dissolved in it) and probably derive and utilize in their nutrition some organic substances manufactured by the gametophytes. Liverwort spores are meiospores; i.e., they arise by meiosis from cells called sporocytes.

The sporophytes may consist almost completely of fertile (sporogenous) tissues (Riccia, Oxymitra), or they may contain sterile cells (nurse cells or elaters) among the developing spores. In Marchantia and Porella, a sterile foot and seta, or stalk, are present; the foot anchors the spore-bearing capsule (sporangium) to the gametophyte and probably serves an absorptive function. The seta connects the foot and the capsule. The elongation of the seta raises the capsule from its protective envelopes, thus placing it in a favourable position for spore dispersal. The capsules of liverworts may shed their spores only by decay of the capsule wall and gametophytic tissues (Riccia, Oxymitra), or they may open irregularly or into two or four segments. Spore germination in some species may occur immediately after deposition if the spores are in a favourable environment, or, as in other species, the spores may require a period of dormancy before germination.

Mosses

In mosses, as in liverworts and hornworts, the leafy shoots belong to the gametophytic phase and produce sex organs when they mature. The leafy shoots (often called gametophores, because they bear the sex organs) arise from a preliminary phase called the protonema, the direct product of spore germination. Filamentous, straplike, or membranous, it grows along the soil surface. A protonema of a moss may proliferate, apparently indefinitely, under favourable conditions and thus increase the population of leafy shoots that arise as buds. Under adverse conditions, certain buds and branches of the protonema may thicken their walls and thus serve to tide the species over an unfavourable growing period.

Life cycle of a moss.

The antheridia and archegonia may be borne at the tips (apices) of the main shoots or on special lateral branchlets. Both bisexual and unisexual leafy shoots occur, depending on the species. In a number of mosses (Mnium, Polytrichum, Funaria), the sexually mature shoots become recognizable through the production of special prominent leaves that form an apical cup around the sex organs. If the cup is brightly coloured, it is often flowerlike. In species with bisexual leafy gametophores, the archegonia and antheridia may be present on the same apex (as can be seen, for example, in Bryum) or at the apices of separate branches as is exemplified in cord moss (Funaria).

Polytrichum moss (Polytrichum).

The archegonia and antheridia of mosses are large enough in many species to be just visible to the unaided eye. The jacket cells of the antheridia are often coloured bright orange or rust; their sperm are biflagellate. As in liverworts and hornworts, rains and even heavy dews evoke the liberation of sperm and the opening of the mature archegonia so that fertilization may be accomplished.

The moss sporophyte, which is attached to the gametophyte, photosynthesizes during much of its development and is more or less self-supporting. It is, to a certain degree, dependent upon the gametophyte for nutrients such as water and mineral salts and, in some cases, even for elaborated foods.

After elongation of the moss sporophyte has ceased, the distal portion (farthest away) enlarges to form the capsule (sporangium), or spore-bearing region. The spores (meiospores), which arise by meiosis, are shed from the capsule gradually through a variety of mechanisms. After the operculum (cover) of the capsule has been shed, its mouth is usually partially closed by the peristome (teeth) and sometimes by associated structures. These teeth absorb moisture, and their resultant swelling and contraction open spaces through which the spores are shed.

Tracheophyte Reproductive Systems

Spore Plants

on short lateral branches. During development, some of the potentially spore-bearing tissue is used as nutrient by the sporocytes as they complete the meiotic divisions that result in colourless kidney-shaped spores. The latter, which are shed as the sporangia open along three lines, germinate and slowly develop into cylindrical sparingly branched gametophytes about 0.5–2 mm (0.02–0.08 inch) in diameter and several millimetres long. They presumably derive nourishment from decaying matter and occur in humus-rich soil, in rock fissures, or among roots on the trunks of tree ferns. The cells of the gametophyte contain fungal structures (hyphae) that probably are involved in some type of nutritional relation with the gametophyte.

In liverworts, hornworts, and mosses, the dominant phase in the life cycle is the sexual gametophyte. In the tracheophytes (vascular cryptogams [which lack true flowers or seeds] and seed plants), on the other hand, the sporophyte is the dominant phase in the life cycle. The gametophytes of the vascular cryptogams mature after the spores that initiated them have been shed from the parent plant, so the gametophytes are free-living. In the seed plants, the gametophytes mature as parasites on the sporophytes.

Psilopsids

The trilobed sporangia of the whisk fern Psilotum not a true fern but a psilopsid are borne terminally on short lateral branches. During development, some of the potentially spore-bearing tissue is used as nutrient by the sporocytes as they complete the meiotic divisions that result in colourless kidney-shaped spores. The latter, which are shed as the sporangia open along three lines, germinate and slowly develop into cylindrical sparingly branched gametophytes about 0.5–2 mm (0.02–0.08 inch) in diameter and several millimetres long. They presumably derive nourishment from decaying matter and occur in humus-rich soil, in rock fissures, or among roots on the trunks of tree ferns. The cells of the gametophyte contain fungal structures (hyphae) that probably are involved in some type of nutritional relation with the gametophyte.

The gametophytes are bisexual and the sperm cells are multiflagellate. The embryonic sporophyte is not easily distinguished from the gametophyte that bears it. At first anchored to the gametophyte by an absorptive foot, the sporophyte ultimately becomes separated from both the foot and the gametophyte.

Lycopsids

In the genus Lycopodium, the sporangia are closely associated with the leaves. In some species (L. lucidulum), the sporangium-bearing leaves (sporophylls) occur in zones among the vegetative portions of the stems. In most, however, the sporophylls occur in specialized compressed stems called cones or strobili. Each sporophyll is associated with one yellow to orange kidney-shaped sporangium.

In several species the spores develop rapidly on the soil surface into ovoid-cylindrical gametophytes about 2–3 mm (0.08–0.12 inch) long, with green lobes and colourless bases; they usually contain a fungus. In other species, development of a colourless gametophyte is slow, so at maturation, which may require up to eight years, the fleshy gametophyte will have become buried in successive layers of humus. These subterranean gametophytes, which contain fungi, are long-lived and are larger (up to 2 cm [0.8 inch]) than the surface types.

The gametophytes of Lycopodium are bisexual, although the antheridia and archegonia may develop into separate groups. The sperms are biflagellate and apparently more than one egg of the same gametophyte may be fertilized.

The zygote divides at a right angle to the long axis of the archegonium. The inner cell gives rise to the embryo, which thus is oriented as if it will develop within the gametophyte; it turns 180° during later development, however, and the axis grows vertically outward from the gametophyte.

In contrast to Lycopodium, the sporophytes of all spike mosses (Selaginella) have sporophylls localized in strobili, and all species of Selaginella are heterosporous; that is, they produce spores of two sizes, the larger designated as megaspores and the smaller as microspores. The megaspores develop into female gametophytes and the microspores into male gametophytes. Accordingly, strobili bear megasporophylls that contain megasporangia, which will produce megaspores, and microsporophylls that contain microsporangia, which will yield microspores. Although the evolutionary origin of two kinds of spores (dimorphism) is unknown, the development of megaspores in living plants suggests that differences in nutrition in the two kinds of sporangia are significant. In a microsporangium, most of the microsporocytes undergo meiosis, forming four spores each; by contrast, all but one or, occasionally, several of the sporocytes in the megasporangium do not complete development. As a result, only four megaspores usually mature in such a sporangium, enlarging as they become gorged with the nutrients made available by disintegration of the other cells. The megaspores, accordingly, are much larger than microspores, although both contain stored food. Both types of spores are thick-walled, and both have prominent three-part (triradiate) ridges.

Unlike the homosporous spores of most liverworts, hornworts, mosses, ferns, and Lycopodium which are characterized by morphologically identical spores that germinate to produce bisexual (both male and female) or unisexual (either male or female) gametophytes the spores of Selaginella begin to develop into gametophytes before they have been shed from their sporangia and attain maturity on a suitable moist substrate.

The microscopic male gametophyte is composed essentially of a single antheridium, which produces biflagellate sperm. The female gametophyte, which protrudes after the megaspore wall cracks open in the region of the triradiate ridge, consists of vegetative cells, has several archegonia at maturity, and usually has three groups of rhizoids. Both male and female gametophytes lack chlorophyll (green pigment), which is necessary for photosynthesis; they utilize nutrients stored in the spores.

After fertilization, one zygote of each female gametophyte develops into an embryonic sporophyte. There is considerable variation in details of development among the species of Selaginella. In some, the spores may develop mature gametophytes before they are shed from their sporangia, and fertilization may occur, so female gametophytes with embryos may be found in the strobili (compressed stems, or cones). The megaspores of Selaginella, containing female gametophytes with still-attached juvenile sporophytes, have the superficial appearance of germinating seeds, from which, however, they differ in many significant respects.

Isoetes, like Selaginella, is monoecious (or hermaphroditic) and heterosporous. Most of the leaves are fertile; some bear one large megasporangium each, and others support a single microsporangium on the inner surface of a spoonlike leaf base. The microsporangia can produce enormous numbers of microspores as many as 1,000,000 and the megasporangia give rise to 50 to 300 megaspores. The spores are liberated as the older sporophylls decay. Unlike those of Selaginella, the spores of Isoetes do not germinate until they have been shed from their sporangia. The unisexual gametophytes are much like those of Selaginella, but the sperm are multiflagellate. The embryonic sporophyte is nourished by food stored in the megaspore and transported through a massive foot.

Sphenopsids

The perennial sporophytes of horsetails (Equisetum species) produce strobili once during every growing season. They may be borne at the tips of green shoots (E. hyemale, E. kansanum), at the tips of nongreen shoots that become green after the spores have been shed (E. fluviatile, E. sylvaticum), or on special nongreen branches that wither and die after the spores have been shed (E. arvense, E. talmateia). The appendages of the strobilus are often called sporangiophores and have been considered to be both stem branches and of leafy origin; in the latter case, they are called sporophylls. Each sporangiophore bears a number of fingerlike sporangia, which produce large numbers of thin-walled, green spores. The outermost wall layer of the spore breaks down into four appendages, which, by their sensitivity to moisture, coil and uncoil, thereby disseminating the spores.

The spores of Equisetum germinate rapidly and grow into green pincushion-like gametophytes anchored to the surface by rhizoids. Apparently, two types of gametophytes are produced from the homosporous spores; some mature slowly, are smaller than others, and always produce antheridia, never archegonia. Others are larger and hermaphroditic, producing archegonia at first and, later, antheridia. The ratios of male to hermaphroditic gametophytes vary among species but are relatively uniform within a species. The ratios are altered by changes in environmental conditions; for example, at certain temperatures (e.g., 32 °C [about 90 °F]), only male gametophytes develop from the spores of five species, whereas at 15 °C (59 °F), approximately 50 percent are male and 50 percent are hermaphroditic gametophytes.

Self-fertilization of hermaphroditic gametophytes can occur, and several sporophytes may be produced on one gametophyte. The embryo consists of an absorptive foot, a primary root (radicle), and a shoot with whorled appendages.

Ferns

As they mature, many fern sporophytes begin to produce spores in clusters of sporangia on the undersurfaces of their fronds (or vegetative "leaves"). Others produce their sporangia on highly modified leaves or portions thereof.

The life cycle of the fern.

- Clusters (sori) of sporangia (spore cases) grow on the undersurface of mature fern leaves.
- Released from its spore case, the haploid spore is carried to the ground, where it germinates into a tiny, usually heart-shaped, gametophyte (gamete-producing structure), anchored to the ground by rhizoids (rootlike projections).
- Under moist conditions, mature sperm are released from the antheridia and swim to the egg-producing archegonia that have formed on the gametophyte's lower surface.
- When fertilization occurs, a zygote forms and develops into an embryo within the archegonium.
- The embryo eventually grows larger than the gametophyte and becomes a sporophyte.

The site of origin of the sporangia is the receptacle; the latter, with its groups of sporangia, is called a sorus. In many ferns, each sorus is covered with a special outgrowth, the indusium; in others the sporangia are covered during development by the margin of the leaf. In a few ferns (e.g., Polypodium), the sori remain uncovered.

In primitive ferns, such as Ophioglossum and Botrychium, the spores are borne upon a specialized axis, the fertile spike. The sporangia of such primitive ferns are massive, with several layers of cellular walls, and produce an indefinite but large number of spores. In most other ferns, the sporangia are smaller and long-stalked, with single-layered walls and a definite number of spores. The spores of the latter are shed explosively by breakage and shrinking as the sporangia open and then slam shut.

Most ferns produce one kind of spore (homospory), but a few genera of aquatic and amphibious ferns (Marsilea, Salvinia, and Azolla) produce two kinds (heterospory), small microspores and much larger megaspores. In either case, after being shed from the parent sporophyte, the spores that have suitable environmental conditions germinate and develop into the gametophytic phase. The ribbonlike, filamentous or heart-shaped gametophytes of most ferns contain chlorophyll, are anchored to some surface moist soil, moist rocks, or tree bark by unicellular rootlike rhizoids, and rarely exceed 13 mm (0.5 inch) in diameter. In a few ferns (Ophioglossum, Botrychium, and certain species of Schizaea), the gametophytes are subterranean, lack chlorophyll, are cylindrical or tuberous, and contain the filamentous structures (hyphae) of an associated fungus.

Fern gametophytes, often called prothalli (singular, prothallus or prothallium), are one cell layer thick except in the centre. Most fern prothalli are bisexual; i.e., they have both male (antheridia) and female (archegonia) sex organs, which develop usually on the undersurface of the prothallus.

Although the eggs of several archegonia may be fertilized, only one zygote usually develops into a juvenile sporophyte. The latter consists of an absorbing foot; a primary root, or radicle, which promptly penetrates the surface; a prominent first leaf; and a rudimentary, slow-growing stem. As the juvenile sporophyte becomes established, the parental gametophyte dies. The series of leaves formed from the stem of the juvenile sporophyte gradually attain the form and vein pattern that characterize the mature sporophyte.

In most ferns, the antheridia appear before the archegonia and continue to develop as the latter mature; furthermore, the archegonial necks curve toward the mature antheridia so that fertilization can readily occur. Both gametes may be derived from one individual or from different individuals. In the bracken fern (Pteridium aquilinum), although the gametophytes are bisexual, self-incompatibility factors reduce self-fertilization. In Onoclea sensiblis, the gametophytes are unisexual in early development, thus favouring cross-fertilization, but later the gametophytes become bisexual so that, if cross-fertilization fails, the species can still be maintained.

Seed Plants

In the two great groups of seed plants, gymnosperms and angiosperms, the sporophyte is the dominant phase in the life cycle, as it is also in the vascular cryptogams; the gametophytes are microscopic parasites on the sporophytes.

In the gymnosperms, the seeds occur individually, exposed at the ends of stalks, sometimes in whorls on an axis, or on the scales of a cone, or megastrobilus. In angiosperms, or flowering plants, by contrast, the seeds are enclosed during development in a structure variously termed a pistil or a carpel, which is sometimes considered to represent an enfolded megasporophyll.

A number of parts of the reproductive process are common to both angiosperms and gymnosperms:

- They produce seeds at maturity.
- The megasporangium, unlike that of heterosporous seedless plants, is covered by one or two cellular layers called integuments and is termed an ovule.
- There is a minute passageway, or micropyle, through the integuments.
- The ovule matures as a seed.
- Only one megasporocyte is present and undergoes meiosis in the megasporangium to produce four megaspores, only one of which usually is functional.
- The megaspore is never discharged from its megasporangium and ovule.
- One female gametophyte is produced within each megasporangium and ovule.
- The microspores begin their development into male gametophytes while still enclosed in the microsporangia.
- As they mature, the male gametophytes, which are contained within the microspore wall and are termed pollen grains, develop a tube that conveys sperm to the egg cell.
- Union of sperm and egg and development of an embryonic sporophyte from the zygote occur within the female gametophyte (sometimes called the "embryo sac"), which is covered by the remains of the megasporangium and by integuments.
- As the embryo develops, the ovule matures as a seed.

In contrast to this impressive list of similarities are important differences, which, in addition to seed position, serve to distinguish angiosperms from gymnosperms. The reproductive cycle in most angiosperms is completed more quickly than that in gymnosperms, and the gametophytes are smaller and simpler and, unlike those of most gymnosperms, lack archegonia. The pollen in angiosperms is transferred to the surface of the megasporophyll, whereas in gymnosperms it is brought to the micropyle of the ovule itself. Two sperm cells are involved in the sexual union in angiosperms: one unites with the egg to form a zygote; the other unites with two nuclei of the female gametophyte to form the primary endosperm nucleus. The latter divides to form a post-fertilization storage tissue, which serves as a food source for the embryo; the embryo of gymnosperms is nourished by the somatic (nonreproductive) tissues of the female gametophyte. The angiosperm ovule increases to mature seed size after fertilization, whereas in gymnosperms, this enlargement occurs prior to fertilization.

Gymnosperms

The cycads are slow-growing dioecious (species with individuals that are either male or female) gymnosperms, the microsporangia (potential pollen) and megasporangia (potential ovules) occurring on different individual sporophytes. In all cycads except the genus Cycas, the ovules are borne on megasporophylls in megastrobili; in Cycas the ovules develop on individual leaflike megasporophylls in what is regarded as a primitive arrangement. The microspores of all cycads develop into microstrobili.

The microspores reach the three-celled stage of development of the male gametophyte before they are shed as pollen grains from the microsporangia. At this time, elongation of the megastrobilus separates the megasporophylls, and the wind-borne pollen grains have access to the micropyles of the ovules. At the time of pollination, each ovule exudes a mucilaginous droplet, the pollination droplet, through the micropyle; some of the pollen grains become engulfed in this droplet and are drawn into the ovule.

The interval between pollination and fertilization is several months in cycads. The sperm cells are multiflagellate and are among the largest (about 300 μm, or 0.01 inch) in the plant kingdom. Each pollen tube may contain 2–22 sperm cells, depending on the genus. The pollen tubes, which develop from the pollen grains, work their way through the megasporangium of the ovule to the archegonia of the female gametophyte. Fertilization of the eggs of the several archegonia is followed by the early development of several embryos (polyembryony), only one of which survives in the mature seeds. Cycad embryos produce two seed leaves, or cotyledons. The seeds are brightly coloured (yellow or scarlet) and covered by an outer fleshy layer and a stony layer of the integument. The seeds of some cycads (e.g., Cycas) may germinate in the megastrobilus without a period of dormancy.

The maidenhair tree, or ginkgo (Ginkgo biloba), is classified separately in a group of which it is the sole living representative. The mature ginkgo (sporophyte) produces microstrobili and ovules each spring as the buds unfold. They occur on the spur shoots among the bases of the young leaves. The ginkgo, like the cycads, is strictly dioecious, so some trees produce ovules and others produce pollen. The ovules occur in pairs at the tips of stalks that emerge among the leaf bases.

Ginkgo pollen, like that of pines, is four-celled at the time of pollination (spring season), which is accomplished by wind. Development of male and female gametophytes is similar to that in cycads, and the sperm cells are also multiflagellate. The female gametophyte, within the ovule of G. biloba, is unique among seed plants in containing chlorophyll. The ovules enlarge tremendously after pollination, and, as the seeds mature, the integument differentiates into several coats, of which a stony layer and an outer fleshy layer are most prominent. The latter becomes mottled, purplish green, and foul smelling. Its tissues may cause nausea or skin eruptions in humans. The inner tissues of the seed (the embryo and the female gametophyte) are palatable and prized among some peoples. Fertilization often occurs after the ovules have fallen from the trees, three or four months after pollination. The ginkgo embryo has two cotyledons.

The sporophytes of most of the species of living conifers, like those of the ginkgo, are woody trees at maturity. They usually grow for a number of years beyond the seedling stage before they mature and produce seeds.

The sporophyte of a typical conifer, such as a pine, may become a large tree. Unlike the cycads and ginkgo, a pine is monoecious, both microstrobili and megastrobili occurring on the same tree. At the beginning of each growing season, the microstrobili enlarge and emerge from their bud scales; they are borne at the base of the terminal bud, which is destined to develop into the current season's growth. The megastrobili, by contrast, arise singly or in a whorl near the apex of the current season's growth.

The microstrobili are called simple strobili, because the microsporangia are borne in pairs on the appendages (microsporophylls) that emerge from the axis of the strobilus. The megastrobili, however, are compound, for the ovules are borne in pairs upon the upper (adaxial) surface of scales, which, in turn, are borne on bracts attached to the megastrobilus.

The pollen of pine, four-celled when shed, is characterized by two lateral air-filled "wings," enlarged cavities between two layers of the pollen-grain wall. The pollen is produced in large amounts and may be transported great distances by air currents. During the time of pollination, the ovuliferous scales on the megastrobili separate slightly, and pollen can be trapped in the pollination droplet of the micropyles of the ovules. Pollen grains that make contact with a droplet are transferred by its subsequent contraction through the micropyle and to the surface of a small depression (pollen chamber) at the tip of the megasporangium.

As a pollen grain germinates, forming a tube that works its way through the megasporangium, it arrives at the female gametophyte as the latter matures its several archegonia. The pollen tube discharges its sperm nuclei into the archegonia, and fertilization is accomplished. As in the cycads and ginkgo, the zygotes of several archegonia may initiate embryogeny. Furthermore, in pine and certain other conifers, the young embryos may form several embryos. At maturity of the seed, however, only one embryo is normally present, embedded in the remains of the female gametophyte and megasporangium, all surrounded by the seed coat (the former integument).

The reproductive process in pine occupies two full growing seasons: ovules pollinated in the spring of a given year do not mature as seeds until the late summer of the next year. The interval between pollination and fertilization is about 14 months.

Among the numerous other gymnosperm species are many different reproductive processes. Some gymnosperms, for example, are dioecious, with microstrobili and megastrobili being borne on separate plants, as in junipers (Juniperus), plum yews (Cephalotaxus), yews (Taxus), and podocarps (Podocarpus). Furthermore, in larch (Larix) and other groups, the pollen grains lack wings. The pollen grains in larch become attached at pollination to a special receptive enlargement of the integument. In podocarps, the megasporangium bulges through the micropyle at pollination and receives the pollen directly. The interval between pollination and fertilization may be as short as four to five weeks in firs (Abies). The number of ovules formed on the ovuliferous scale varies, as does the number of microsporangia on the microsporophyll. There may be only one ovule in a megastrobilus, as in some junipers, and the megastrobili may become fleshy, also in junipers. In yews the solitary ovules are terminal on dwarf shoots; each ovule is surrounded by a cuplike structure called an aril, which becomes fleshy and brightly coloured as the seed matures. The number of sperm produced in each male gametophyte varies also from 2 in pine to 20 in some cypresses (Cupressus).

The genera Ephedra, Gnetum, and Welwitschia, which are often grouped together in one category (Gnetales, or Gnetophyta), differ among themselves and from other gymnosperms with respect to several details of reproduction. The microsporangia and ovules of both Ephedra and Welwitschia are produced in compound strobili; those of Gnetum are borne in a series of whorls on elongated axes sometimes misleadingly called "inflorescences". The ovules of these genera, unlike those of other gymnosperms, have two integuments instead of one, as in angiospermous ovules. Archegonia are present in the female gametophytes of Ephedra, but only eggs occur in those of Gnetum and Welwitschia. The sperm, like those of the conifers, lack flagella.

Angiosperms

Although the angiosperms are known as flowering plants, they are difficult to distinguish from gymnosperms solely on the basis of bearing flowers, for, like the strobilus, a flower is a compressed stem, with crowded spore-bearing appendages. The occurrence of coloured petals and attractive scents is not essential and is by no means characteristic of all flowers. The most important distinguishing feature separating flowering plants from gymnosperms is that the ovules of flowering plants are produced within enclosed containers called carpels.

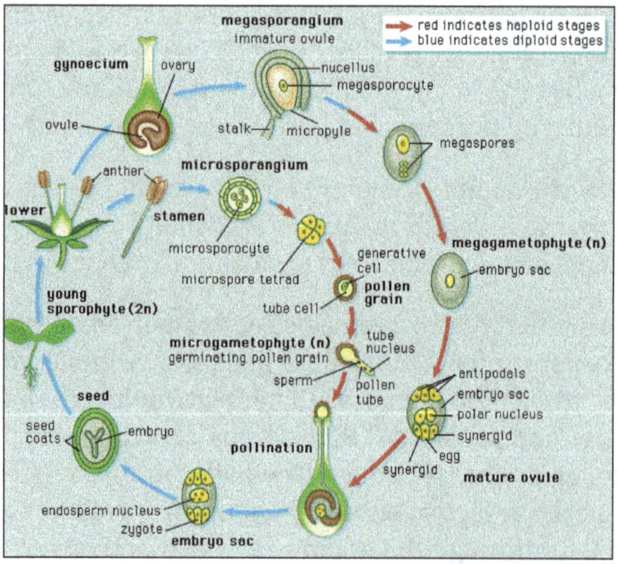

Life Cycle of a Typical Angiosperm

The angiosperm life cycle consists of a sporophyte phase and a gametophyte phase. The cells of a sporophyte body have a full complement of chromosomes (i.e., the cells are diploid, or 2n); the sporophyte is the typical plant body that one sees when one looks at an angiosperm. The gametophyte arises when cells of the sporophyte, in preparation for reproduction, undergo meiotic division and produce reproductive cells that have only half the number of chromosomes (i.e., haploid, or n). A two-celled microgametophyte (called a pollen grain) germinates into a pollen tube and through division produces the haploid sperm. An eight-celled megagametophyte (called the embryo sac) produces the egg. Fertilization occurs with the fusion of a sperm with an egg to produce a zygote, which eventually develops into an embryo. After fertilization, the ovule develops into a seed, and the ovary develops into a fruit.

Flowers may occur singly at the ends of stems (e.g., tulip, poppy, rose), or they may be grouped in various clusters, or inflorescences (gladiolus, sunflower, delphinium, and yarrow).

Tulip (tulipa).

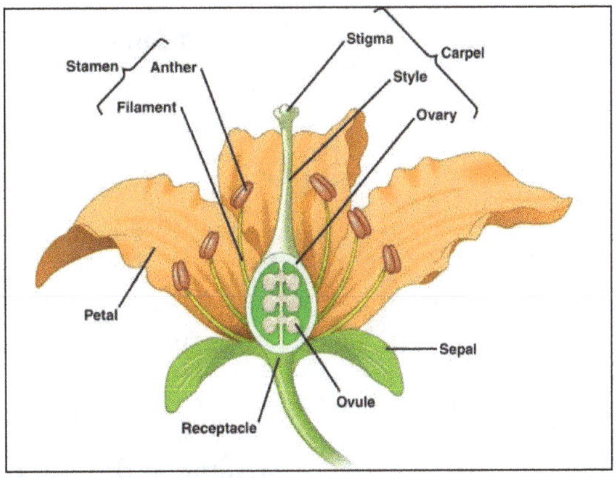

Diagram of a typical flowering plant (angiosperm).

An individual flower may be complete, in that a given floral receptacle produces sepals (often greenish and leaflike), petals (often white or coloured other than green), stamens, and a pistil (or pistils). The sepals are collectively known as the calyx, and the petals as the corolla; the calyx and corolla compose the perianth. If sepals or petals are lacking, the flower is said to be incomplete. Although incomplete, a flower that has both stamens and a pistil is said to be perfect; lacking either of these parts, it is imperfect.

In practice, groups of solitary flowers are not easily distinguished from inflorescences; the latter seemingly evolved from a system of branches, each with a terminal solitary flower. The inflorescence may be few flowered or have up to 6,000,000 flowers, as in certain palms. Inflorescences vary also in their position, being terminal, axillary, or intercalary. Terminal inflorescences are at the tips of the major, or dominant, branches; axillary ones are at the tips of axillary, or side, branches. In intercalary inflorescences, the stem continues beyond the inflorescence, which may result in alternating fertile and sterile areas of the axis.

Inflorescences can be distinguished by their growth patterns as determinate or indeterminate. In determinate inflorescences the first formed flower at the tip of the dominant stem matures first, and younger flowers develop on lower lateral branches; the cyme of the forget-me-not (Myosotis) is a typical example. In indeterminate inflorescences the growing region of the axis functions for extended periods so that as the older flowers mature and set fruit near the base of the inflorescence axis, younger buds develop and continue to expand into flowers at the apex. This is exemplified in the spikes of yucca and the racemes of delphinium, in which the youngest flowers are farthest away from the root. Other types of indeterminate inflorescences include umbels and capitula, or heads. The youngest flower is terminal or central in umbels and in heads.

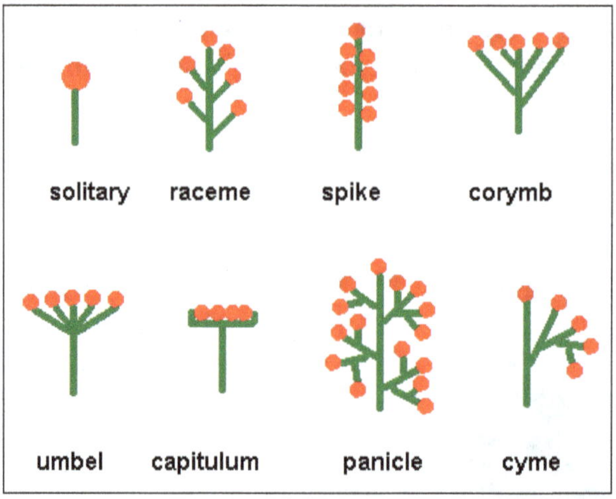

Angiosperm inflorescences: Common types of inflorescences among the angiosperms.

The head is the type of inflorescence that characterizes the Asteraceae, or aster family. It may be few to many flowered and usually has at its base one or more series of leaflike bracts. The small individual flowers arise in spiral order on the receptacle, the youngest being at the centre. The basal calyx of each flower, known as a pappus, is bristlelike, scaly, or feathery and borne at the top of the ovary. The corolla, formed of the petals, may be (1) tubular, with five petal lobes, sometimes split open, (2) ligulate, or tongue-like, with a very short basal tube, or (3) bilabiate, with the tube split into two tips. In some genera, all the flowers are ligulate, whereas in others, the marginal flowers are ligulate (ray flowers) and the others tubular or all are tubular. The marginal ray flowers are either female (pistillate) or sterile. The tubular flowers are characterized by male and female parts: five united pollen-bearing stamens and a pistil, which matures as a one-seeded fruit (achene).

The position of the floral organs with reference to each other and to the tip of the floral receptacle varies in different flowers; in some, the perianth (sepals and petals) and stamens are attached to the receptacle below the pistil; such flowers are hypogynous (e.g., buttercup and magnolia). In others (rose, cherry, peach), the perianth and stamens are borne on the rim of a concave structure in the depression of which the pistil is borne; such flowers are perigynous (i.e., borne on a ring or cup of the receptacle surrounding a pistil). Finally, there are flowers in which the ovary is enclosed by a tissue composed of the fused bases of the perianth and stamens (apple, pear, aster); the blossom seems to arise upon or above the ovary and is called epigynous (i.e., appearing to grow from the top of the ovary).

The stamen, seemingly the equivalent of the gymnospermous microsporophyll, consists of an anther (a group of two to four microsporangia) borne at the tip of a blade stalk, or filament. The pistil, most often composed of an enlarged basal ovary, a columnar style, and distal stigma, is the ovule-producing organ of the flower. It is often considered to have evolved from enfolded megasporophyll or some other ovuliferous structure with enclosed ovules (angiospermy); alternatively, it is thought to have arisen from the cuplike bracts of extinct seed-bearing plants on which the leafy bracts grew together and thus enclosed the ovules.

There may be one or more pistils on the floral receptacle, depending on the species. Furthermore, pistils may be simple (composed of one ovule-bearing unit, megasporophyll, or carpel) or compound (composed of more than one carpel). Compound pistils are thought to have arisen as a result of crowding of simple pistils on the floral axis; for example, variation in the degree of fusion may be observed in members of the saxifrage family. The ovary which matures as the fruit usually reveals by the number of ovule-containing chambers (locules) the number of carpels it contains. The stigma is a specially adapted portion of the pistil modified for the reception of pollen. It may be feathery and branched or elongated, as in such wind-pollinated flowers as those of the grasses, or it may be compact and have a sticky surface. The ovary may contain one ovule (e.g., buckwheat, avocado), a few ovules (e.g., grape, bean) or a large number of ovules (tobacco, begonia, snapdragon).

In some angiosperms (e.g., corn, hickory, walnut, pecan, oak), both types of imperfect flower are borne on the same plant, which is therefore called monoecious. By contrast, staminate flowers may occur on one plant and pistillate flowers on another, as in willows, poplars, and mulberries, which are dioecious. In common parlance, staminate flowers and plants that bear them are often designated "male," and pistillate flowers and the plants that bear them are called "female". This may be traced back at least as far as to the time of Swedish botanist Carolus Linnaeus, who interpreted stamens and pistils as sex organs. Comparative morphology indicates clearly, however, that stamens and pistils are the spore-bearing structures of the sporophyte and not actually the gamete-bearing organs of the gametophyte. The terms "male" and "female," applied to angiosperm plants and their flowers, is often condoned because the gametophytic phase is so condensed in angiosperms. The designations suggest to the uninitiated, however, that pollen grains and sperm, on the one hand, and eggs and ovules, on the other, are identical, which is not the case.

Among the vast number of species of angiosperms, there is considerable variation in floral organization. The perianth may be absent or present; it may be clearly differentiated as calyx and corolla (e.g., pea); or the perianth segments may be similar (magnolia, tulip tree). The number of stamens and pistils may be large and separately attached to the receptacle in a spiral pattern (buttercup), or the numbers may be reduced and the attachment cyclic or whorled (lily). The stamens may be fused by their anthers (daisy) or their filaments (peas, beans). The filaments may be petal-like (water lilies) or stalk-like. Opening of the anther may be by longitudinal or transverse fissures or by terminal pores.

The reproductive cycle in angiosperms can be traced from before the shedding of pollen. The microspores begin their development of male gametophytes, which involves formation of a small generative cell and a tube cell. The generative cell may divide to form two sperm cells before the pollen grain (developing male gametophyte) is shed or while the pollen tube is growing during germination. The pollen grains of angiosperms have variously, and often elaborately, ornamented walls characteristic of the species.

The pistil of a flower may receive pollen from the stamens of the same flower, in self-pollination (e.g., peas and tomatoes). In many other flowers, however, pollen from one or more flowers is transferred to the stigmas of other flowers. A number of specialized relationships have evolved between floral organization and animal pollinators such as insects. In the majority of angiosperms, one megasporocyte develops in the megasporangium (often called the nucellus) of the ovule, and a tetrad of megaspores is formed as a result of meiosis. Three megaspores (nearest the micropyle) degenerate; only one enlarges, and then it undergoes divisions to form the eight-nucleate, seven-celled female gametophyte ("embryo sac"). Of the three cells of this gametophyte near the micropyle, one functions as an egg. As the pollen tube discharges its contents into the female gametophyte, the egg nucleus is fertilized by one of the sperm cells, and the other unites with the two nuclei (polar nuclei) within the large central cell of the female gametophyte. The resultant nucleus, which has three sets of chromosomes, is the primary endosperm nucleus. This process, double fertilization, occurs only in angiosperms.

Both pollination and fertilization stimulate cell division in the ovary, ovules, and zygotes, all of which enter upon a period of rapid enlargement. In most angiosperms, the primary endosperm nucleus divides to form endosperm tissue, the cells of which become filled with stored food, such as starches, oils, and proteins. As the rate of embryonic development decreases, the seeds of most angiosperms enter a period of dormancy, accompanied by dehydration and hardening of the integuments, which form seed coats. At this period, the enlarged ovary (and sometimes adjacent structures) matures as fruit. Angiosperm seeds may germinate as soon as they reach maturity, or they may undergo various kinds of dormancy.

Some representative variations occur in the reproductive process of angiosperms. In violets (Viola), in addition to the ordinary flowers produced first during the usual flowering season, less conspicuous flowers later develop; called cleistogamous flowers, they do not open but are self-pollinated, thus ensuring augmentation of the population during a period less favourable for the usual blossoms.

The pollen grains of most angiosperms separate from each other, but in some cases (e.g., Rhododendron), they remain attached in original groups of four, called tetrads. The very tiny pollen grains of orchids, certain mimosas, and milkweeds are clustered in waxy masses called pollinia (singular pollinium).

A number of variations in pattern of development of the female gametophyte occur in various angiosperms; for example, in certain species of evening primrose (Oenothera), the female gametophyte contains only 4 nuclei, whereas in Peperomia, as many as 16 may be present. In lily, all 4 megaspore nuclei are involved in the formation of the female gametophyte.

Pollen may germinate immediately after contact with a stigma (sugarcane), within five minutes (corn), in two hours (beet), or after one or two days. The pollen grains of most plants produce only one pollen tube, but 10 or more pollen tubes have been observed to develop from one pollen grain in plants of the mallow family. The pollen tubes usually enter through the micropyle (porogamy), but they may also enter through the base of the ovule (chalazogamy).

The interval between pollination and fertilization varies. It may be as long as 12–14 months in certain species of oak, 5–7 months in witch hazel, 2–20 weeks among the orchids, 3–4 hours in lettuce, and as little as 15–45 minutes in dandelions.

The postfertilization endosperm fails to develop in orchid seeds but is present at least during early embryogeny in most others. The endosperm may arise by nuclear divisions and become cellular as nuclear divisions terminate, or its development may involve both nuclear and cell divisions from the beginning. In a number of cases (e.g., legumes), the embryo consumes the endosperm during its development, resulting in a mature seed with a massive embryo and no endosperm. Most angiosperm embryos have two seed leaves (are dicotyledonous); some have one lateral cotyledon (are monocotyledonous); and a few (e.g., Degeneria) have three to four cotyledons.

In seed germination, the cotyledons may remain below the soil surface within the seed (hypogean germination) and may function in digesting and absorbing endosperm (corn). In addition, some may serve as sources of stored food themselves (pea). Other cotyledons may rise above the soil surface (epigean germination) by elongation of the hypocotyl, the embryonic axis between the root and the growing stem, or epicotyl. Cotyledons that emerge above the soil may wither and drop off as their food is used (e.g., bean), or they may persist and function as photosynthetic leaves (e.g., castor bean).

An even greater range of variation occurs in angiospermous fruits. The fruit may arise from one pistil (simple or compound) of one flower (e.g., the simple fruits of pea and peach), from several pistils of one flower (e.g., the aggregate fruits of strawberry and raspberry), or from the pistils of several flowers (e.g., the multiple fruits of pineapple, mulberry, and corn). Simple fruits may be dry (legumes) or fleshy (peach, apple, tomato) at maturity. Dry fruits may open (dehisce; many legumes) or remain closed about the seed (be indehiscent; grasses and sunflower).

The manner of ovular attachment is known as placentation. The ovary may contain one to many ovules, which may be attached to the ovary wall (parietal placentation) or to the central axis (axial, or free-central, placentation). Despite these and other variations in the morphology of flower parts, the reproductive process is, with minor diversities, remarkably uniform.

Variations in Reproductive Cycles

The life cycles and reproductive processes characterize the vast majority of their respective plant groups.

Bracken (Pteridium aquilinum).

Among the liverworts it has been demonstrated that small fragments of the stalk of the sporophyte are capable of regenerating diploid gametophytes. In the mosses, both haploid and diploid apospory have been experimentally evoked. As in the liverworts, injury and regeneration of fragments of the sporophytic seta result in diploid gametophytes. By contrast, fragments of moss leaves, stems, and rhizoids (and even the sterile tissues of the sex organs) can regenerate haploid gametophytes.

In certain strains of mosses, the gametophyte can give rise to clusters of presumably haploid sporophytes without the functioning of gametes; such apogamous formation of sporophytes may also be chemically induced (by application of a solution containing a specific amount of chloral hydrate to both the protonema and leafy shoots).

Among the vascular plants, both natural and induced apogamy and apospory are known. In certain ferns, gametophytes may develop at the leaf margins or in sori from transformed sporangia. Certain other ferns reproduce apogamously in nature; thus, for example, in the holly fern (Crytomium falcatum), the gametophytes give rise directly to sporophytes by nuclear and cell division on vegetative cells of the gametophyte. In almost every group, however, variations of the usual reproductive process occur. These may involve substitution of asexual reproduction for sexual or the direct production of plants by cells other than the usual ones (apomixis). Apomictic phenomena which are in the strictest sense asexual include apospory, in which the gametophyte phase is produced without the need of spores, and apogamy, in which the sporophyte phase is produced without the need of gametes, or sex cells.

Apogamy may be induced in normally sexual ferns by withholding water from the gametophytes, which prevents the liberation and functioning of sperm. Similarly, when gametophytes are grown in inorganic culture media supplemented by a variety of sugars, they produce sporophytes apogamously. Colourless roots removed from the bracken fern (Pteridium aquilinum) have been induced to develop diploid gametophytes aposporously, as have the injured juvenile leaves of a number of ferns.

Apomictic phenomena occur also among many angiosperms. In some species, haploid sporophytes may develop either from the unfertilized egg or from some other cell of the gametophyte. Such apogamy occurs, for example, after stimulation of one species with the pollen of a related one (e.g., Solanum nigrum by the pollen of S. luteum). Apogamy involving an unfertilized egg (a phenomenon termed parthenogenesis) occurs in certain orchids. Male parthenogenesis, or the production of a sporophyte from a sperm, has been detected in tobacco hybrids. Finally, a form of haploid apogamy is known in which a cell of the female gametophyte other than an egg may develop into an embryo.

In certain species of hawkweed, the embryo develops from a certain cell of the ovule or the megasporangium. In others, the female gametophyte is diploid through an impairment of the meiotic process; in this case, the egg (diploid parthenogenesis) or one of the related cells may form an embryo. In citrus trees a number of embryos (polyembryony) arise from diploid cells of the megasporangium or integuments.

Physiology of Plant Reproduction

The maturation of sporophytes and gametophytes, as manifested by their ability to produce spores and gametes, respectively, involves both internal and environmental factors. With respect to the former, the organism must have completed a certain minimum period of vegetative development before environmental factors are able to stimulate formation of spores and gametes.

Marchantia: The liverwort Marchantia polymorpha.

Among environmental factors affecting reproduction, the duration, intensity, and quality of light, as well as temperature, have primary roles; for example, the liverwort Marchantia polymorpha continues in the vegetative state indefinitely under daily fluorescent illumination of 16 hours. Control plants exposed to daily incandescent lighting of 16 hours become sexually mature after 30 days. Addition of sucrose hastens the development of the sexual phase, an indication that chemical factors also play a role. In hornworts (e.g., Anthoceros and Phaeoceros), antheridia develop under daily light periods of 4–12 hours; none develop when the plants are illuminated for longer periods.

Temperature also affects sexual maturation. In mosses, for example, initiation of sex organs in bacteria-free laboratory cultures of Funaria occurs at 10°C (50°F) when cultures are illuminated 6, 12, or 20 hours daily. On the other hand, the moss Polytrichum is seemingly not affected by duration of light but forms sex organs best at 21° C (70° F).

Among vascular cryptogams little is known about the various environmental factors that affect development of sex organs in nature, and, except for certain ferns and horsetails, experimental studies of controlled laboratory cultures are lacking. It has been shown that laboratory cultures of gametophytes of certain horsetails derived from single spores produce sex organs 40–60 days after the spores germinate. In horsetails generally, conditions favouring vegetative growth evoke a preponderance of initially female gametophytes; less favourable conditions induce the production of male gametophytes.

Considerable information is available about the physiology of reproduction in ferns, especially with respect to the gametophyte generation. Spore germination occurs if adequate moisture is present and in temperatures between 15 and 35°C (59 and 95°F). The spores germinate and the young gametophytes do best under neutral or slightly acid conditions. Most fern spores require light for germination. Some fern spores remain viable for as long as 20 years.

Spores germinating in darkness or in the absence of blue wavelengths of light remain in the protonemal, or filamentous, condition. Growth to the mature gametophyte requires light of blue wavelengths in most species.

An interesting aspect of fern reproductive physiology is the discovery that antheridial production is under hormonal control. Several types of antheridium-evoking substances (hormones) have been recognized, and there is some evidence that they are specific. The sperm of ferns are attracted chemically to the vicinity of the maturing archegonia, where they become trapped in extruded mucilage. Although only one sporophyte usually develops on a gametophyte, as many as five may be induced to form if the gametophyte is repeatedly exposed to sperm.

The preceding paragraphs, dealing with vascular cryptogams, have emphasized the influence of environmental factors on the expression of sexual reproduction; the remaining account deals with the angiosperms. Because the gametophytic phase of angiosperms is abbreviated and because it is parasitic on the sporophyte, the maturation of the sporophyte, as manifested by flowering, has received more intensive study in angiosperms than has that of the gametophyte.

Angiosperms that complete their life cycles within one growing season (in the temperate zone) are known as annuals. Perhaps the shortest angiospermous life cycle known is that of Plantago insularis, a southern California species, that can grow from seed to the production of its own seed in four to six weeks under domestication; the cycle of most annuals, however, is longer. A number of angiosperms are biennial in the temperate zone: they grow vegetatively for one season, but their flowering and seed production are delayed until a second growing season, after which the plants die (e.g., beets, carrots). Still other angiosperms are perennial: they continue growing, flowering, and producing seeds for a number of growing seasons (e.g., irises, roses, oaks, eucalypti).

It has been demonstrated that duration and wavelength of light are of paramount importance in controlling the flowering of angiosperms. On the basis of extensive experimental studies, flowering plants have been classified as "long-day," "short-day," or "day-neutral" with respect to their requirements of light for flowering. What they actually perceive is the length of the night, using their phytochrome system, which employs blue pigments in photoperiodic processes. Botanists use flashes of red or "far-red" light to interrupt the dark period, which allows them to classify various plants with respect to their photoperiodism.

Temperature also plays an important role in flowering. Thus, members of the cabbage family can be grown without flowering when high enough temperatures are maintained. A number of perennial and biennial plants (bulbous plants, beets) require a prior period of low temperature before they flower. In addition to internal (i.e., genetic factors), therefore, temperature and light are of paramount importance for many plants in evoking maturation.

Several other physiological aspects of reproduction in angiosperms are noteworthy. Once transported to the stigma, the germination of the pollen grains is markedly affected by the chemical composition of the stigmatic exudate; e.g., pollen of an unrelated species will not germinate on the stigma. There is evidence also that the directional growth of the pollen tubes through the style toward the ovular micropyle is stimulated by a chemical substance produced by the style, ovules, and probably other tissues of the pistil.

It is also clear that both pollination and fertilization have physiological effects on fruit production. In a number of cases, pollination alone (not followed by fertilization) is sufficient stimulus to evoke enlargement of the pistil to form a seedless (parthenocarpic) fruit. This phenomenon occurs in banana and in certain varieties of citrus fruits, grapes, and cucumbers. Parthenocarpy can also be induced by exposure of the stigma to indoleacetic acid, naphthoxyacetic acid, indole butyric acid, naphthaleneacetic acid, or other synthetic hormones. The hormone gibberellin is effective in producing seedless grapes and is the active component in preparations used to prevent premature dropping of fruit.

Also significant in the reproduction of flowering plants are the phenomena of self-sterility and self-incompatibility, which prevail in certain plants with perfect flowers (having both stamens and pistils). Pollen from one flower or from another flower of the same plant or from a plant with identical genetic constitution, when applied to the stigma, fails to germinate or grows so slowly that fertilization does not occur. For example, sweet cherries are self-incompatible; a number of groups of genetically intrasterile types are known, but cross-pollination of trees of different genetic groups results in fruit production. Self-incompatibility occurs also in certain strains of garden plants and in plum, apple, pear, and tobacco.

Self-sterility, usually associated with differences in chromosome number, often occurs in hybrids from widely divergent parents. This is true of cultivated varieties of blackberry and raspberry. Such chromosomal imbalance creates irregularities in meiosis during formation of the pollen, which results in infertility.

PLANT REPRODUCTIVE MORPHOLOGY

Plant reproductive morphology is the study of the physical form and structure (the morphology) of those parts of plants directly or indirectly concerned with sexual reproduction.

Close-up of a flower of Schlumbergera (Christmas or Holiday Cactus), showing part of the gynoecium (the stigma and part of the style is visible) and the stamens that surround it.

Among all living organisms, flowers, which are the reproductive structures of angiosperms, are the most varied physically and show a correspondingly great diversity in methods of reproduction. Plants that are not flowering plants (green algae, mosses, liverworts, hornworts, ferns and gymnosperms such as conifers) also have complex interplays between morphological adaptation and environmental factors in their sexual reproduction. The breeding system, or how the sperm from one plant fertilizes the ovum of another, depends on the reproductive morphology, and is the single most important determinant of the genetic structure of nonclonal plant populations. Christian Konrad Sprengel studied the reproduction of flowering plants and for the first time it was understood that the pollination process involved both biotic and abiotic interactions. Charles Darwin's theories of natural selection utilized this work to build his theory of evolution, which includes analysis of the coevolution of flowers and their insect pollinators.

Dioicous gametophytes of the liverwort Marchantia polymorpha. In this species, gametes are produced on different plants on umbrella-shaped gametophores with different morphologies. The radiating arms of female gameteophores (left) protect archegonia that produce eggs. Male gametophores (right) are topped with antheridia that produce sperm.

Plants have complex lifecycles involving alternation of generations. One generation, the sporophyte, gives rise to the next generation asexually via spores. Spores may be identical isospores or come in different sizes (microspores and megaspores), but strictly speaking, spores and sporophytes are neither male nor female because they do not produce gametes. The alternate generation, the gametophyte, produces gametes, eggs and/or sperm. A gametophyte can be monoicous (bisexual), producing both eggs and sperm or dioicous (unisexual), either female (producing eggs) or male (producing sperm).

In the bryophytes (liverworts, mosses and hornworts), the sexual gametophyte is the dominant generation. In ferns and seed plants (including cycads, conifers, flowering plants, etc.) the sporophyte is the dominant generation. The obvious visible plant, whether a small herb or a large tree, is the sporophyte, and the gametophyte is very small. In seed plants, each female gametophyte, and the spore that gives rise to it, is hidden within the sporophyte and is entirely dependent on it for nutrition. Each male gametophyte typically consists of from two to four cells enclosed within the protective wall of a pollen grain.

The sporophyte of a flowering plant is often described using sexual terms (e.g. "female" or "male") based on the sexuality of the gametophyte it gives rise to. For example, a sporophyte that produces spores that give rise only to male gametophytes may be described as "male", even though the sporophyte itself is asexual, producing only spores. Similarly, flowers produced by the sporophyte may be described as "unisexual" or "bisexual", meaning that they give rise to either one sex of gametophyte or both sexes of gametophyte.

Flowering Plants

Basic Flower Morphology

The flower is the characteristic structure concerned with sexual reproduction in flowering plants (angiosperms). Flowers vary enormously in their construction (morphology). A "complete" flower, like that of Ranunculus glaberrimus, has a calyx of outer sepals and a corolla of inner petals. The sepals and petals together form the perianth. Next inwards there are numerous stamens, which produce pollen grains, each containing a microscopic male gametophyte. Stamens may be called the "male" parts of a flower and collectively form the androecium. Finally in the middle there are carpels, which at maturity contain one or more ovules, and within each ovule is a tiny female gametophyte. Carpels may be called the "female" parts of a flower and collectively form the gynoecium.

Flower of Ranunculus glaberrimus.

Each carpel in Ranunculus species is an achene that produces one ovule, which when fertilized becomes a seed. If the carpel contains more than one seed, as in Eranthis hyemalis, it is called a follicle. Two or more carpels may be fused together to varying degrees and the entire structure, including the fused styles and stigmas may be called a pistil. The lower part of the pistil, where the ovules are produced, is called the ovary. It may be divided into chambers (locules) corresponding to the separate carpels.

Variations

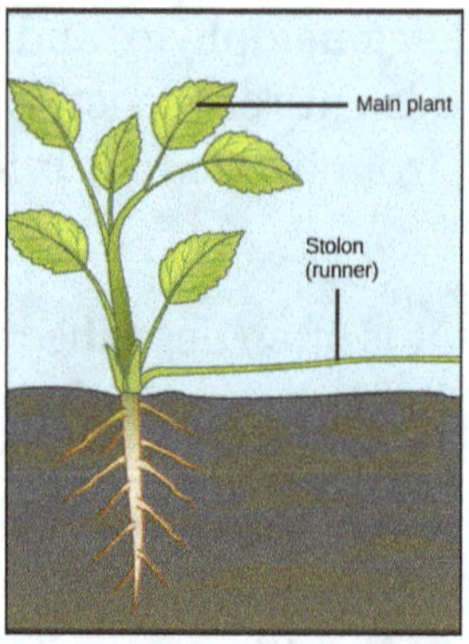

Alnus serrulata has unisexual flowers and is monoecious. Maturing male flower catkins on the right, last year's female catkins on the left.

Ilex aquifolium is dioecious: (above) shoot with flowers from male plant; (top right) male flower enlarged, showing stamens with pollen and reduced, sterile stigma; (below) shoot with flowers from female plant; (lower right) female flower enlarged, showing stigma and reduced, sterile stamens (staminodes) with no pollen.

A "perfect" flower has both stamens and carpels, and may be described as "bisexual" or "hermaphroditic". A "unisexual" flower is one in which either the stamens or the carpels are missing, vestigial or otherwise non-functional. Each flower is either "staminate" (having only functional stamens) and thus "male", or "carpellate" (or "pistillate") (having only functional carpels) and thus "female". If separate staminate and carpellate flowers are always found on the same plant, the species is called monoecious. If separate staminate and carpellate flowers are always found on different plants, the species is called dioecious. A 1995 study found that about 6% of angiosperm species are dioecious, and that 7% of genera contain some dioecious species.

Members of the birch family (Betulaceae) are examples of monoecious plants with unisexual flowers. A mature alder tree (Alnus species) produces long catkins containing only male flowers, each with four stamens and a minute perianth, and separate stalked groups of female flowers, each without a perianth.

Most hollies (members of the genus Ilex) are dioecious. Each plant produces either functionally male flowers or functionally female flowers. In Ilex aquifolium, the common European holly, both kinds of flower have four sepals and four white petals; male flowers have four stamens, female flowers usually have four non-functional reduced stamens and a four-celled ovary. Since only female plants are able to set fruit and produce berries, this has consequences for gardeners. Amborella represents the first known group of flowering plants to separate from their common ancestor. It too is dioecious; at any one time, each plant produces either flowers with functional stamens but no carpels, or flowers with a few non-functional stamens and a number of fully functional carpels. However, Amborella plants may change their "sex" over time. In one study, five cuttings from a male plant produced only male flowers when they first flowered, but at their second flowering three switched to producing female flowers.

In extreme cases, all of the parts present in a complete flower may be missing, so long as at least one carpel or one stamen is present. This situation is reached in the female flowers of duckweeds (Lemna), which comprise a single carpel, and in the male flowers of spurges (Euphorbia) which comprise a single stamen.

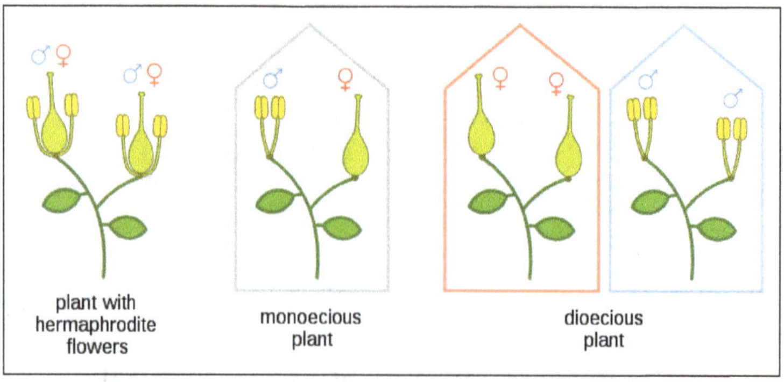

The basic cases of sexuality of flowering plants.

A species such as Fraxinus excelsior, the common ash of Europe, demonstrates one possible kind of variation. Ash flowers are wind-pollinated and lack petals and sepals. Structurally, the flowers may be bisexual, consisting of two stamens and an ovary, or may be male (staminate), lacking a functional ovary, or female (carpellate), lacking functional stamens. Different forms may occur on the same tree, or on different trees. The Asteraceae (sunflower family), with close to 22,000 species worldwide, have highly modified inflorescences made up of flowers (florets) collected together into tightly packed heads. Heads may have florets of one sexual morphology – all bisexual, all carpellate or all staminate (when they are called homogamous), or may have mixtures of two or more sexual forms (heterogamous). Thus goatsbeards (Tragopogon species) have heads of bisexual florets, like other members of the tribe Cichorieae, whereas marigolds (Calendula species) generally have heads with the outer florets bisexual and the inner florets staminate (male).

Like Amborella, some plants undergo sex-switching. For example, Arisaema triphyllum (Jack-in-the-pulpit) expresses sexual differences at different stages of growth: smaller plants produce all or mostly male flowers; as plants grow larger over the years the male flowers are replaced by more female flowers on the same plant. Arisaema triphyllum thus covers a multitude of sexual conditions in its lifetime: nonsexual juvenile plants, young plants that are all male, larger plants with a mix of both male and female flowers, and large plants that have mostly female flowers. Other plant populations have plants that produce more male flowers early in the year and as plants bloom later in the growing season they produce more female flowers.

Outcrossing

Outcrossing, cross-fertilization or allogamy, in which offspring are formed by the fusion of the gametes of two different plants, is the most common mode of reproduction among higher plants. About 55% of higher plant species reproduce in this way. An additional 7% are partially cross-fertilizing and partially self-fertilizing (autogamy). About

15% produce gametes but are principally self-fertilizing with significant out-crossing lacking. Only about 8% of higher plant species reproduce exclusively by non-sexual means. These include plants that reproduce vegetatively by runners or bulbils, or which produce seeds without embryo fertilization (apomixis). The selective advantage of out-crossing appears to be the masking of deleterious recessive mutations.

The primary mechanism used by flowering plants to ensure outcrossing involves a genetic mechanism known as self-incompatibility. Various aspects of floral morphology promote allogamy. In plants with bisexual flowers, the anthers and carpels may mature at different times, plants being protandrous (with the anthers maturing first) or protogynous (with the carpels mature first). Monoecious species, with unisexual flowers on the same plant, may produce male and female flowers at different times.

Dioecy, the condition of having unisexual flowers on different plants, necessarily results in outcrossing, and might thus be thought to have evolved for this purpose. However, "dioecy has proven difficult to explain simply as an outbreeding mechanism in plants that lack self-incompatibility". Resource-allocation constraints may be important in the evolution of dioecy, for example, with wind-pollination, separate male flowers arranged in a catkin that vibrates in the wind may provide better pollen dispersal. In climbing plants, rapid upward growth may be essential, and resource allocation to fruit production may be incompatible with rapid growth, thus giving an advantage to delayed production of female flowers. Dioecy has evolved separately in many different lineages, and monoecy in the plant lineage correlates with the evolution of dioecy, suggesting that dioecy can evolve more readily from plants that already produce separate male and female flowers.

ASEXUAL REPRODUCTION IN PLANTS

Plants can reproduce asexually, without the fertilization of gametes, by either vegetative reproduction or apomixis.

Many plants are able to propagate themselves using asexual reproduction. This method does not require the investment required to produce a flower, attract pollinators, or find a means of seed dispersal. Asexual reproduction produces plants that are genetically identical to the parent plant because no mixing of male and female gametes takes place. Traditionally, these plants survive well under stable environmental conditions when compared with plants produced from sexual reproduction because they carry genes identical to those of their parents.

Plants have two main types of asexual reproduction: vegetative reproduction and apomixis. Vegetative reproduction results in new plant individuals without the production of seeds or spores. Many different types of roots exhibit vegetative reproduction. The corm is used by gladiolus and garlic. Bulbs, such as a scaly bulb in lilies and a tunicate bulb in daffodils, are other common examples of this type of reproduction. A potato is a stem tuber, while parsnip propagates from a taproot. Ginger and iris produce rhizomes, while ivy uses an adventitious root (a root arising from a plant part other than the main or primary root), and the strawberry plant has a stolon, which is also called a runner.

Roots: Different types of stems allow for asexual reproduction. (a) The corm of a garlic plant looks similar to (b) a tulip bulb, but the corm is solid tissue, while the bulb consists of layers of modified leaves that surround an underground stem. Both corms and bulbs can self-propagate, giving rise to new plants. (c) Ginger forms masses of stems called rhizomes that can give rise to multiple plants. (d) Potato plants form fleshy stem tubers. Each eye in the stem tuber can give rise to a new plant. (e) Strawberry plants form stolons: stems that grow at the soil surface or just below ground and can give rise to new plants.

Some plants can produce seeds without fertilization. Either the ovule or part of the ovary, which is diploid in nature, gives rise to a new seed. This method of reproduction is known as apomixis.

An advantage of asexual reproduction is that the resulting plant will reach maturity faster. Since the new plant is arising from an adult plant or plant parts, it will also be sturdier than a seedling. Asexual reproduction can take place by natural or artificial (assisted by humans) means.

Natural and Artificial Methods of Asexual Reproduction in Plants

Plants can undergo natural methods of asexual reproduction, performed by the plant itself, or artificial methods, aided by humans.

Natural Methods of Asexual Reproduction

Natural methods of asexual reproduction include strategies that plants have developed to self-propagate. Many plants, such as ginger, onion, gladioli, and dahlia, continue to grow from buds that are present on the surface of the stem. In some plants, such as the sweet potato, adventitious roots or runners (stolons) can give rise to new plants. In Bryophyllum and kalanchoe, the leaves have small buds on their margins. When these are detached from the plant, they grow into independent plants; they may also start growing into independent plants if the leaf touches the soil. Some plants can be propagated through cuttings alone.

Artificial Methods of Asexual Reproduction

Artificial methods of asexual reproduction are frequently employed to give rise to new, and sometimes novel, plants. They include grafting, cutting, layering, and micropropagation.

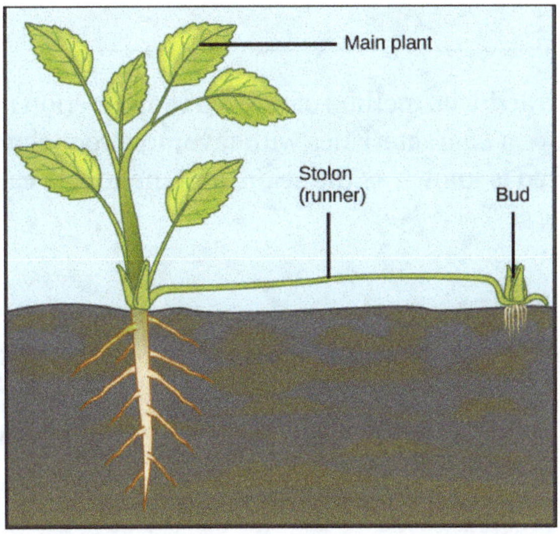

Runners: asexual reproduction: A stolon, or runner, is a stem that runs along the ground. At the nodes, it forms adventitious roots and buds that grow into a new plant.

Grafting

Grafting has long been used to produce novel varieties of roses, citrus species, and other plants. In grafting, two plant species are used: part of the stem of the desirable plant is grafted onto a rooted plant called the stock. The part that is grafted or attached is called the scion. Both are cut at an oblique angle (any angle other than a right angle), placed in close contact with each other, and are then held together. Matching up these two surfaces as closely as possible is extremely important because these will be holding the plant together. The vascular systems of the two plants grow and fuse, forming a graft. After a period of time, the scion starts producing shoots, eventually bearing flowers and fruits. Grafting is widely used in viticulture (grape growing) and the citrus industry. Scions capable of producing a particular fruit variety are grafted onto root stock with specific resistance to disease.

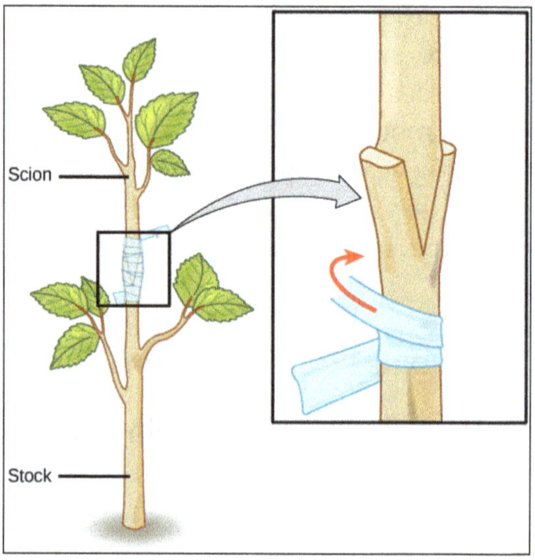

Grafting: Grafting is an artificial method of asexual reproduction used to produce plants combining favorable stem characteristics with favorable root characteristics. The stem of the plant to be grafted is known as the scion, and the root is called the stock.

Cutting

Plants such as coleus and money plant are propagated through stem cuttings where a portion of the stem containing nodes and internodes is placed in moist soil and allowed to root. In some species, stems can start producing a root even when placed only in water. For example, leaves of the African violet will root if kept undisturbed in water for several weeks.

Layering

Layering is a method in which a stem attached to the plant is bent and covered with soil. Young stems that can be bent easily without any injury are the preferred plant for this method. Jasmine and bougainvillea (paper flower) can be propagated this way. In some plants, a modified form of layering known as air layering is employed. A portion of the bark or outermost covering of the stem is removed and covered with moss, which is then taped. Some gardeners also apply rooting hormone. After some time, roots will appear; this portion of the plant can be removed and transplanted into a separate pot.

Micropropagation

Micropropagation (also called plant tissue culture) is a method of propagating a large number of plants from a single plant in a short time under laboratory conditions. This method allows propagation of rare, endangered species that may be difficult to grow under natural conditions, are economically important, or are in demand as disease-free plants.

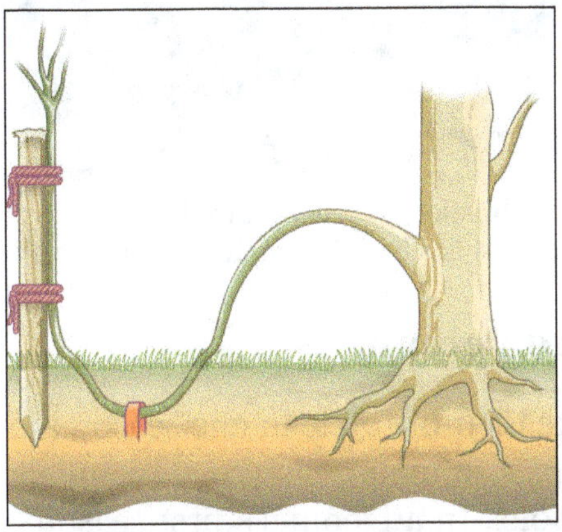

Layering: In layering, a part of the stem is buried so that it forms a new plant.

To start plant tissue culture, a part of the plant such as a stem, leaf, embryo, anther, or seed can be used. The plant material is thoroughly sterilized using a combination of chemical treatments standardized for that species. Under sterile conditions, the plant material is placed on a plant tissue culture medium that contains all the minerals, vitamins, and hormones required by the plant. The plant part often gives rise to an undifferentiated mass, known as a callus, from which, after a period of time, individual plantlets begin to grow. These can be separated; they are first grown under greenhouse conditions before they are moved to field conditions.

Plant Life Spans

The life cycles and life spans of plants vary and are affected by environmental and genetic factors.

The length of time from the beginning of development to the death of a plant is called its life span. The life cycle, on the other hand, is the sequence of stages a plant goes through from seed germination to seed production of the mature plant. Some plants, such as annuals, only need a few weeks to grow, produce seeds, and die. Other plants, such as the bristlecone pine, live for thousands of years. Some bristlecone pines have a documented age of 4,500 years. Even as some parts of a plant, such as regions containing meristematic tissue (the area of active plant growth consisting of undifferentiated cells capable of cell division) continue to grow, some parts undergo programmed cell death (apoptosis). The cork found on stems and the water-conducting tissue of the xylem, for example, are composed of dead cells.

Plant life spans: The bristlecone pine, shown here in the Ancient Bristlecone Pine Forest has been known to live for 4,500 years.

Annuals, Biennial and Perennials

Plant species that complete their life cycle in one season are known as annuals, an example of which is Arabidopsis, or mouse-ear cress. Biennials, such as carrots, complete their life cycle in two seasons. In a biennial's first season, the plant has a vegetative phase, whereas in the next season, it completes its reproductive phase. Commercial growers harvest the carrot roots after the first year of growth and do not allow the plants to flower. Perennials, such as the magnolia, complete their life cycle in two years or more.

Monocarpic and Polycarpic Plants

In another classification based on flowering frequency, monocarpic plants flower only once in their lifetime; examples of monocarpic plants include bamboo and yucca. During the vegetative period of their life cycle (which may be as long as 120 years in some bamboo species), these plants may reproduce asexually, accumulating a great deal of food material that will be required during their once-in-a-lifetime flowering and setting of seed after fertilization. Soon after flowering, these plants die. Polycarpic plants form flowers many times during their lifetime. Fruit trees, such as apple and orange trees, are polycarpic; they flower every year. Other polycarpic species, such as perennials, flower several times during their life span, but not each year. By this method, the plant does not require all its nutrients to be channeled towards flowering each year.

Genetics and Environmental Conditions

As is the case with all living organisms, genetics and environmental conditions have a role to play in determining how long a plant will live. Susceptibility to disease, changing environmental conditions, drought, cold, and competition for nutrients are some of the factors that determine the survival of a plant. Plants continue to grow, despite the presence of dead tissue, such as cork. Individual parts of plants, such as flowers and leaves, have different rates of survival. In many trees, the older leaves turn yellow and eventually fall from the tree. Leaf fall is triggered by factors such as a decrease in photosynthetic efficiency due to shading by upper leaves or oxidative damage incurred as a result of photosynthetic reactions. The components of the part to be shed are recycled by the plant for use in other processes, such as development of seed and storage. This process is known as nutrient recycling. However, the complex pathways of nutrient recycling within a plant are not well understood.

The aging of a plant and all the associated processes is known as senescence, which is marked by several complex biochemical changes. One of the characteristics of senescence is the breakdown of chloroplasts, which is characterized by the yellowing of leaves. The chloroplasts contain components of photosynthetic machinery, such as membranes and proteins. Chloroplasts also contain DNA. The proteins, lipids, and nucleic acids are broken down by specific enzymes into smaller molecules and salvaged by the plant to support the growth of other plant tissues. Hormones are known to play a role in senescence. Applications of cytokinins and ethylene delay or prevent senescence; in contrast, abscissic acid causes premature onset of senescence.

Plant senescence: The autumn color of these Oregon Grape leaves is an example of programmed plant senescence.

SEXUAL REPRODUCTION IN PLANTS

There are several different methods and processes involved in the sexual reproduction of plants.

Self-pollination and Cross-pollination

In angiosperms, pollination is defined as the placement or transfer of pollen from the anther to the stigma of the same flower or another flower. In gymnosperms, pollination involves pollen transfer from the male cone to the female cone. Upon transfer, the pollen germinates to form the pollen tube and the sperm for fertilizing the egg. Pollination has been well studied since the time of Gregor Mendel. Mendel successfully carried out self- as well as cross-pollination in garden peas while studying how characteristics were passed on from one generation to the next. Today's crops are a result of plant breeding, which employs artificial selection to produce the present-day cultivars. A case in point is today's corn, which is a result of years of breeding that started with its ancestor, teosinte. The teosinte that the ancient Mayans originally began cultivating had tiny seeds vastly different from today's relatively giant ears of corn. Interestingly, though these two plants appear to be entirely different, the genetic difference between them is miniscule.

Pollination takes two forms: self-pollination and cross-pollination. Self-pollination occurs when the pollen from the anther is deposited on the stigma of the same flower, or another flower on the same plant. Cross-pollination is the transfer of pollen from the anther of one flower to the stigma of another flower on a different individual of the same species. Self-pollination occurs in flowers where the stamen and carpel mature at the same time, and are positioned so that the pollen can land on the flower's stigma. This method of pollination does not require an investment from the plant to provide nectar and pollen as food for pollinators.

Living species are designed to ensure survival of their progeny; those that fail become extinct. Genetic diversity is therefore required so that in changing environmental or stress conditions, some of the progeny can survive. Self-pollination leads to the production of plants with less genetic diversity, since genetic material from the same plant is used to form gametes, and eventually, the zygote. In contrast, cross-pollination or out-crossing leads to greater genetic diversity because the micro-gametophyte and mega-gametophyte are derived from different plants.

Because cross-pollination allows for more genetic diversity, plants have developed many ways to avoid self-pollination. In some species, the pollen and the ovary mature at different times. These flowers make self-pollination nearly impossible. By the time pollen matures and has been shed, the stigma of this flower is mature and can only be pollinated by pollen from another flower. Some flowers have developed physical features that prevent self-pollination. The primrose is one such flower. Primroses have evolved two flower types with differences in anther and stigma length: the pin-eyed flower has anthers positioned at the pollen tube's halfway point, and the thrum-eyed flower's stigma is likewise located at the halfway point. Insects easily cross-pollinate while seeking the nectar at the bottom of the pollen tube. This phenomenon is also known as heterostyly. Many plants, such as cucumber, have male and female flowers located on different parts of the plant, thus making self-pollination difficult. In yet other species, the male and female flowers are borne on different plants (dioecious). All of these are barriers to self-pollination; therefore, the plants depend on pollinators to transfer pollen. The majority of pollinators are biotic agents such as insects (like bees, flies, and butterflies), bats, birds, and other animals. Other plant species are pollinated by abiotic agents, such as wind and water.

Methods of Pollination

Pollination by Insects

Insects, such as bees, are important agents of pollination.

Bees are perhaps the most important pollinator of many garden plants and most commercial fruit trees. The most common species of bees are bumblebees and honeybees. Since bees cannot see the color red, bee-pollinated flowers usually have shades of blue, yellow, or other colors. Bees collect energy-rich pollen or nectar for their survival and energy needs. They visit flowers that are open during the day, are brightly colored, have a strong aroma or scent, and have a tubular shape, typically with the presence of a nectar guide. A nectar guide includes regions on the flower petals that are visible only to bees, and not to humans; it helps to guide bees to the center of the flower, thus making the pollination process more efficient. The pollen sticks to the bees' fuzzy hair, and when the bee visits another flower, some of the pollen is transferred to the second flower. Recently, there have been many reports about the declining population of honeybees. Many flowers will remain unpollinated and not bear seed if honeybees disappear. The impact on commercial fruit growers could be devastating.

Many flies are attracted to flowers that have a decaying smell or an odor of rotting flesh. These flowers, which produce nectar, usually have dull colors, such as brown or purple. They are found on the corpse flower or voodoo lily (Amorphophallus), dragon arum (Dracunculus), and carrion flower (Stapleia, Rafflesia). The nectar provides energy, whereas the pollen provides protein. Wasps are also important insect pollinators, and pollinate many species of figs.

A corn earworm sips nectar from a night-blooming Gaura plant.

Butterflies, such as the monarch, pollinate many garden flowers and wildflowers, which usually occur in clusters. These flowers are brightly colored, have a strong fragrance, are open during the day, and have nectar guides to make access to nectar easier. The pollen is picked up and carried on the butterfly's limbs. Moths, on the other hand, pollinate flowers during the late afternoon and night. The flowers pollinated by moths are pale or white and are flat, enabling the moths to land. One well-studied example of a moth-pollinated plant is the yucca plant, which is pollinated by the yucca moth. The shape of the flower and moth have adapted in such a way as to allow successful pollination. The moth deposits pollen on the sticky stigma for fertilization to occur later. The female moth also deposits eggs into the ovary. As the eggs develop into larvae, they obtain food from the flower and developing seeds. Thus, both the insect and flower benefit from each other in this symbiotic relationship. The corn earworm moth and Gaura plant have a similar relationship.

Pollination by Bats

In the tropics and deserts, bats are often the pollinators of nocturnal flowers such as agave, guava, and morning glory. The flowers are usually large and white or pale-colored; thus, they can be distinguished from the dark surroundings at night. The flowers have a strong, fruity, or musky fragrance and produce large amounts of nectar. They are naturally large and wide-mouthed to accommodate the head of the bat. As the bats seek the nectar, their faces and heads become covered with pollen, which is then transferred to the next flower.

Pollination by Birds

Hummingbirds have adaptations that allow them to reach the nectar of certain tubular flowers.

Many species of small birds, such as the hummingbird and sun birds, are pollinators for plants such as orchids and other wildflowers. Flowers visited by birds are usually sturdy and are oriented in such a way as to allow the birds to stay near the flower without getting their wings entangled in the nearby flowers. The flower typically has a curved, tubular shape, which allows access for the bird's beak. Brightly colored, odorless flowers that are open during the day are pollinated by birds. As a bird seeks energy-rich nectar, pollen is deposited on the bird's head and neck and is then transferred to the next flower it visits. Botanists have been known to determine the range of extinct plants by collecting and identifying pollen from 200-year-old bird specimens from the same site.

Pollination by Wind

A person knocks pollen from a pine tree.

Most species of conifers, and many angiosperms, such as grasses, maples and oaks, are pollinated by wind. Pine cones are brown and unscented, while the flowers of wind-pollinated angiosperm species are usually green, small, may have small or no petals, and produce large amounts of pollen. Unlike the typical insect-pollinated flowers, flowers adapted to pollination by wind do not produce nectar or scent. In wind-pollinated species, the microsporangia hang out of the flower, and, as the wind blows, the lightweight pollen is carried with it.

The flowers usually emerge early in the spring, before the leaves, so that the leaves do not block the movement of the wind. The pollen is deposited on the exposed feathery stigma of the flower.

These male (a) and female (b) catkins are from the goat willow tree (Salix caprea). Note how both structures are light and feathery to better disperse and catch the wind-blown pollen.

Pollination by Water

Some weeds, such as Australian sea grass and pond weeds, are pollinated by water. The pollen floats on water, and when it comes into contact with the flower, it is deposited inside the flower.

Pollination by Deception

Certain orchids use food deception or sexual deception to attract pollinators. Shown here is a bee orchid (Ophrys apifera).

Orchids are highly valued flowers, with many rare varieties. They grow in a range of specific habitats, mainly in the tropics of Asia, South America, and Central America. At least 25,000 species of orchids have been identified.

Flowers often attract pollinators with food rewards, in the form of nectar. However, some species of orchid are an exception to this standard: they have evolved different ways to attract the desired pollinators. They use a method known as food deception, in which bright colors and perfumes are offered, but no food. Anacamptis morio, commonly known as the green-winged orchid, bears bright purple flowers and emits a strong scent. The bumblebee, its main pollinator, is attracted to the flower because of the strong scent which usually indicates food for a bee and in the process, picks up the pollen to be transported to another flower.

Other orchids use sexual deception. Chiloglottis trapeziformis emits a compound that smells the same as the pheromone emitted by a female wasp to attract male wasps. The male wasp is attracted to the scent, lands on the orchid flower, and in the process, transfers pollen. Some orchids, like the Australian hammer orchid, use scent as well as visual trickery in yet another sexual deception strategy to attract wasps. The flower of this orchid mimics the appearance of a female wasp and emits a pheromone. The male wasp tries to mate with what appears to be a female wasp, and in the process, picks up pollen, which it then transfers to the next counterfeit mate.

Double Fertilization

After pollen is deposited on the stigma, it must germinate and grow through the style to reach the ovule. The microspores, or the pollen, contain two cells: the pollen tube cell and the generative cell. The pollen tube cell grows into a pollen tube through which the generative cell travels. The germination of the pollen tube requires water, oxygen, and certain chemical signals. As it travels through the style to reach the embryo sac, the pollen tube's growth is supported by the tissues of the style. In the meantime, if the generative cell has not already split into two cells, it now divides to form two sperm cells. The pollen tube is guided by the chemicals secreted by the synergids present in the embryo sac, and it enters the ovule sac through the micropyle. Of the two sperm cells, one sperm fertilizes the egg cell, forming a diploid zygote; the other sperm fuses with the two polar nuclei, forming a triploid cell that develops into the endosperm. Together, these two fertilization events in angiosperms are known as double fertilization. After fertilization is complete, no other sperm can enter. The fertilized ovule forms the seed, whereas the tissues of the ovary become the fruit, usually enveloping the seed.

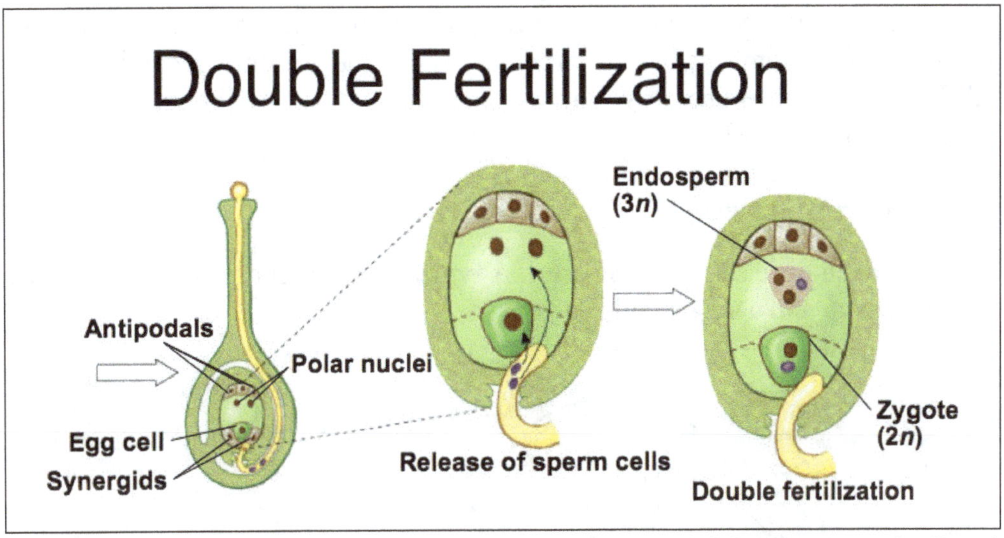

In angiosperms, one sperm fertilizes the egg to form the 2n zygote, and the other sperm fertilizes the central cell to form the 3n endosperm. This is called a double fertilization.

After fertilization, the zygote divides to form two cells: the upper cell, or terminal cell, and the lower, or basal, cell. The division of the basal cell gives rise to the suspensor, which eventually makes connection with the maternal tissue. The suspensor provides a route for nutrition to be transported from the mother plant to the growing embryo. The terminal cell also divides, giving rise to a globular-shaped proembryo. In dicots (eudicots), the developing embryo has a heart shape, due to the presence of the two rudimentary cotyledons. In non-endospermic dicots, such as Capsella bursa, the endosperm develops initially, but is then digested, and the food reserves are moved into the two cotyledons. As the embryo and cotyledons enlarge, they run out of room inside the developing seed, and are forced to bend (Figure 8c). Ultimately, the embryo and cotyledons fill the seed, and the seed is ready for dispersal. Embryonic development is suspended after some time, and growth is resumed only when the seed germinates. The developing seedling will rely on the food reserves stored in the cotyledons until the first set of leaves begin photosynthesis.

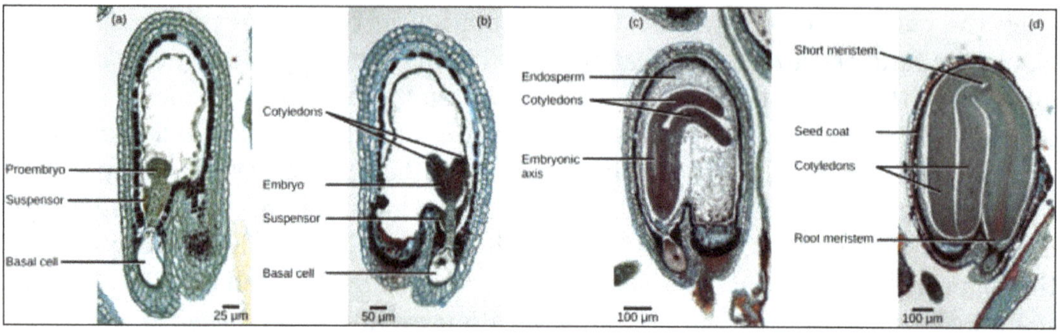

The stages of embryo development in the ovule of a shepherd's purse (Capsella bursa): after fertilization, the zygote divides to form an upper terminal cell and a lower basal cell. (a) In the first stage of development, the terminal cell divides, forming a globular pro-embryo. The basal cell also divides, giving rise to the suspensor. (b) In the second stage, the developing embryo has a heart shape due to the presence of cotyledons. (c) In the third stage, the growing embryo runs out of room and starts to bend. (d) Eventually, it completely fills the seed.

Development of a Seed

The mature ovule develops into the seed. A typical seed contains a seed coat, cotyledons, endosperm, and a single embryo.

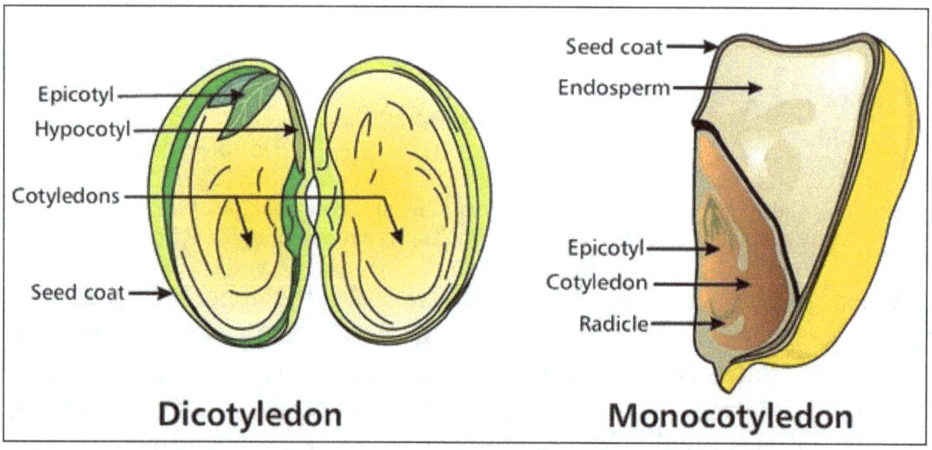

The structures of dicot and monocot seeds are shown. Dicots (left) have two cotyledons. Monocots, such as corn (right), have one cotyledon, called the scutellum; it channels nutrition to the growing embryo. Both monocot and dicot embryos have a plumule that forms the leaves, a hypocotyl that forms the stem, and a radicle that forms the root.

The storage of food reserves in angiosperm seeds differs between monocots and dicots. In monocots, such as corn and wheat, the single cotyledon is called a scutellum; the scutellum is connected directly to the embryo via vascular tissue (xylem and phloem).

Food reserves are stored in the large endosperm. Upon germination, enzymes are secreted by the aleurone, a single layer of cells just inside the seed coat that surrounds the endosperm and embryo. The enzymes degrade the stored carbohydrates, proteins and lipids, the products of which are absorbed by the scutellum and transported via a vasculature strand to the developing embryo. Therefore, the scutellum can be seen to be an absorptive organ, not a storage organ.

The two cotyledons in the dicot seed also have vascular connections to the embryo. In endospermic dicots, the food reserves are stored in the endosperm. During germination, the two cotyledons therefore act as absorptive organs to take up the enzymatically released food reserves, much like in monocots (monocots, by definition, also have endospermic seeds). Tobacco (Nicotiana tabaccum), tomato (Solanum lycopersicum), and pepper (Capsicum annuum) are examples of endospermic dicots. In non-endospermic dicots, the triploid endosperm develops normally following double fertilization, but the endosperm food reserves are quickly remobilized and moved into the developing cotyledon for storage. The two halves of a peanut seed (Arachis hypogaea) and the split peas (Pisum sativum) of split pea soup are individual cotyledons loaded with food reserves.

The seed, along with the ovule, is protected by a seed coat that is formed from the integuments of the ovule sac. In dicots, the seed coat is further divided into an outer coat known as the testa and inner coat known as the tegmen.

The embryonic axis consists of three parts: the plumule, the radicle, and the hypocotyl. The portion of the embryo between the cotyledon attachment point and the radicle is known as the hypocotyl (hypocotyl means "below the cotyledons"). The embryonic axis terminates in a radicle (the embryonic root), which is the region from which the root will develop. In dicots, the hypocotyls extend above ground, giving rise to the stem of the plant. In monocots, the hypocotyl does not show above ground because monocots do not exhibit stem elongation. The part of the embryonic axis that projects above the cotyledons is known as the epicotyl. The plumule is composed of the epicotyl, young leaves, and the shoot apical meristem.

Upon germination in dicot seeds, the epicotyl is shaped like a hook with the plumule pointing downwards. This shape is called the plumule hook, and it persists as long as germination proceeds in the dark. Therefore, as the epicotyl pushes through the tough and abrasive soil, the plumule is protected from damage. Upon exposure to light, the hypocotyl hook straightens out, the young foliage leaves face the sun and expand, and the epicotyl continues to elongate. During this time, the radicle is also growing and producing the primary root. As it grows downward to form the tap root, lateral roots branch off to all sides, producing the typical dicot tap root system.

In monocot seeds, the testa and tegmen of the seed coat are fused. As the seed germinates, the primary root emerges, protected by the root-tip covering: the coleorhiza. Next, the primary shoot emerges, protected by the coleoptile: the covering of the shoot tip. Upon exposure to light (i.e. when the plumule has exited the soil and the protective coleoptile is no longer needed), elongation of the coleoptile ceases and the leaves expand and unfold. At the other end of the embryonic axis, the primary root soon dies, while other, adventitious roots (roots that do not arise from the usual place – i.e. the root) emerge from the base of the stem. This gives the monocot a fibrous root system.

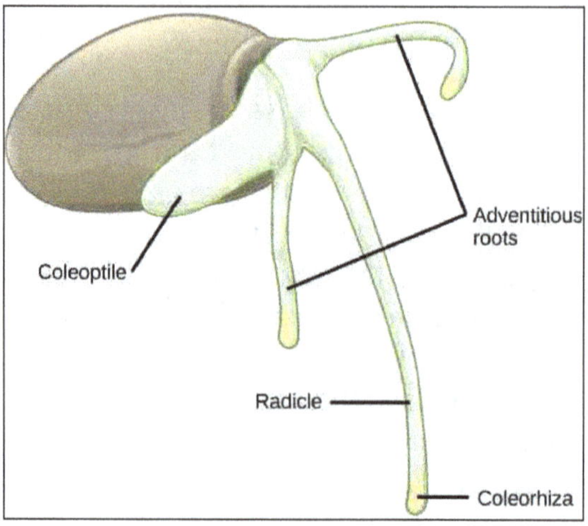

As this monocot grass seed germinates, the primary root, or radicle, emerges first, followed by the primary shoot, or coleoptile, and the adventitious roots.

Seed Germination

Many mature seeds enter a period of inactivity, or extremely low metabolic activity: a process known as dormancy, which may last for months, years or even centuries. Dormancy helps keep seeds viable during unfavorable conditions. Upon a return to favorable conditions, seed germination takes place. Favorable conditions could be as diverse as moisture, light, cold, fire, or chemical treatments. After heavy rains, many new seedlings emerge. Forest fires also lead to the emergence of new seedlings. Some seeds require vernalization (cold treatment) before they can germinate. This guarantees that seeds produced by plants in temperate climates will not germinate until the spring. Plants growing in hot climates may have seeds that need a heat treatment in order to germinate, to avoid germination in the hot, dry summers. In many seeds, the presence of a thick seed coat retards the ability to germinate. Scarification, which includes mechanical or chemical processes to soften the seed coat, is often employed before germination. Presoaking in hot water, or passing through an acid environment, such as an animal's digestive tract, may also be employed.

Depending on seed size, the time taken for a seedling to emerge may vary. Species with large seeds have enough food reserves to germinate deep below ground, and still extend their epicotyl all the way to the soil surface. Seeds of small-seeded species usually require light as a germination cue. This ensures the seeds only germinate at or near the soil surface (where the light is greatest). If they were to germinate too far underneath the surface, the developing seedling would not have enough food reserves to reach the sunlight.

Development of Fruit and Fruit Types

After fertilization, the ovary of the flower usually develops into the fruit. Fruits are usually associated with having a sweet taste; however, not all fruits are sweet. Botanically, the term "fruit" is used for a ripened ovary. In most cases, flowers in which fertilization has taken place will develop into fruits, and flowers in which fertilization has not taken place will not. Some fruits develop from the ovary and are known as true fruits, whereas others develop from other parts of the female gametophyte and are known as accessory fruits. The fruit encloses the seeds and the developing embryo, thereby providing it with protection. Fruits are of many types, depending on their origin and texture. The sweet tissue of the blackberry, the red flesh of the tomato, the shell of the peanut, and the hull of corn (the tough, thin part that gets stuck in your teeth when you eat popcorn) are all fruits. As the fruit matures, the seeds also mature.

Fruits may be classified as simple, aggregate, multiple, or accessory, depending on their origin. If the fruit develops from a single carpel or fused carpels of a single ovary, it is known as a simple fruit, as seen in nuts and beans. An aggregate fruit is one that develops from more than one carpel, but all are in the same flower: the mature carpels fuse together to form the entire fruit, as seen in the raspberry. Multiple fruit develops from an inflorescence or a cluster of flowers. An example is the pineapple, where the flowers fuse together to form the fruit. Accessory fruits (sometimes called false fruits) are not derived from the ovary, but from another part of the flower, such as the receptacle (strawberry) or the hypanthium (apples and pears).

Simple fruit | Aggregate fruits | Multiple fruits | Accessory fruit

There are four main types of fruits. Simple fruits, such as these nuts, are derived from a single ovary. Aggregate fruits, like raspberries, form from many carpels that fuse together. Multiple fruits, such as pineapple, form from a cluster of flowers called an inflorescence. Accessory fruit, like the apple, are formed from a part of the plant other than the ovary.

Fruits generally have three parts: the exocarp (the outermost skin or covering), the mesocarp (middle part of the fruit), and the endocarp (the inner part of the fruit). Together, all three are known as the pericarp. The mesocarp is usually the fleshy, edible part of the fruit; however, in some fruits, such as the almond, the endocarp is the edible part. In many fruits, two or all three of the layers are fused, and are indistinguishable at maturity. Fruits can be dry or fleshy. Furthermore, fruits can be divided into dehiscent or indehiscent types. Dehiscent fruits, such as peas, readily release their seeds, while indehiscent fruits, like peaches, rely on decay to release their seeds.

Fruit and Seed Dispersal

The fruit has a single purpose: seed dispersal. Seeds contained within fruits need to be dispersed far from the mother plant, so they may find favorable and less competitive conditions in which to germinate and grow.

Some fruit have built-in mechanisms so they can disperse by themselves, whereas others require the help of agents like wind, water, and animals. Modifications in seed structure, composition, and size help in dispersal. Wind-dispersed fruit are lightweight and may have wing-like appendages that allow them to be carried by the wind. Some have a parachute-like structure to keep them afloat. Some fruits for example, the dandelion have hairy, weightless structures that are suited to dispersal by wind.

(a) (b) (c)

Fruits and seeds are dispersed by various means. (a) Dandelion seeds are dispersed by wind, the (b) coconut seed is dispersed by water, and the (c) acorn is dispersed by animals that cache and then forget it.

Seeds dispersed by water are contained in light and buoyant fruit, giving them the ability to float. Coconuts are well known for their ability to float on water to reach land where they can germinate. Similarly, willow and silver birches produce lightweight fruit that can float on water.

Animals and birds eat fruits, and the seeds that are not digested are excreted in their droppings some distance away. Some animals, like squirrels, bury seed-containing fruits for later use; if the squirrel does not find its stash of fruit, and if conditions are favorable, the seeds germinate. Some fruits, like the cocklebur, have hooks or sticky structures that stick to an animal's coat and are then transported to another place. Humans also play a big role in dispersing seeds when they carry fruits to new places and throw away the inedible part that contains the seeds.

MONOCOTYLEDON REPRODUCTION

The monocots (or Monocotyledons) are one of the two major clades of flowering plants (or Angiosperms), the other being the dicots (or dicotyldons). In order to reproduce they utilize various strategies such as employing forms of asexual reproduction, restricting which individuals they are sexually compatible with, or influencing how they are pollinated. Nearly all reproductive strategies that evolved in the dicots have independently evolved in monocots as well. Despite these similarities and their close relatedness, monocots and dicots have distinct traits in their reproductive biologies.

A solitary bee pollinating an Allium monocot flower.

Most monocots reproduce sexually through use of seeds that have a single cotyledon, however a great number of monocots reproduce asexually through clonal propagation. Breeding systems that utilize self-incompatibility are much more common than those that utilize self-compatibility. The majority of monocots are animal pollinated (zoophilous), of which most are pollinator generalists. Monocots have mechanisms to promote or suppress cross-fertilization (allogamy) and self-fertilization (autogamy or geitonogamy). The pollination syndromes of monocots can be quite distinct; they include having flower parts in multiples of three, adaptations to pollination by water (hydrogamy), and pollination by sexual deception in orchids.

Seed Production

Reproducing through seeds is the most widespread method of reproduction in both monocots and dicots. However, internal seed structure is vastly different between these groups. The cotyledon is the embryonic leaf within a seed; monocots have one whereas dicots have two. The evolution of having one or two cotyledons may have arisen 200-150 Mya when monocots and dicots are thought to have diverged. Furthermore, the cotyledons in dicot seeds contain the endosperm which acts as the seed's food storage, while in monocot the endosperm is separated from the cotyledon. Reproduction through seeds is normally a sexual mode of reproduction, however in some cases individuals can asexually produce fertile seeds without pollination, termed apomixis.

Clonal Propagation

Multiple individuals have sprouted from turmeric rhizomes.

Some monocots can reproduce asexually without the need for seeds. Clonal propagation is the production or division of vegetative structures which develop into new individual that are genetically identical to their progenitor. These vegetative structures can also form enlarged tubers that function as food storage. Monocots constitute the majority of plants with such structures, mainly in the families: Iridaceae, Liliaceae and Amaryllidaceae. There are many different types of clonal propagation, which are classified by the type of tissue propagating.

- Rhizomes are root-like stems which usually grow laterally underground and sprout new individuals. Most plants that produce rhizomes are monocots (grasses, bamboo, ginger, galangal, turmeric, orchids, irises, lotus); these include the families: Poaceae, Zingiberaceae, Orchidaceae, Iridaceae, and Nelumbonaceae.
- Stolons (runners) are like rhizomes, except they may grow above ground, are not root-like, and always grow laterally. They can be founded in grasses, irises, and orchids; these include the families: Poaceae, Iridaceae, and Orchidaceae.
- Bulbs are underground food storage structures made from leaves which cycle through periods of vegetative and reproductive growth. Nearly all bulb producing plants are monocots (onion, lily, tulip, hyacinth, irises); these include the families: Amaryllidaceae, Liliaceae, Asparagaceae, and Iridaceae.
- Corms are underground swollen stems that act as food storage; they appear similar to bulbs but are not layered with leaves. New corms will bud around the base of the stem. Corms can be found in irises, taro, arrowheads, sedges, and bananas; these include: Iridaceae, Araceae, Alismataceae, Asparagaceae, Colchicaceae, Cyperaceae, and Musaceae.
- Keikis are clonal individuals that grow from the flowering stems of Orchids.

Breeding Systems

Monocots can be classified as perfect (having bisexual flowers), monoecious (having separate male and female flowers on the same plant), dioecious (having flowers of only one sex on an individual) and polygamous (having bisexual flowers with male and/or female flowers on the same plant). Plants that are dioecious have no other option but to mate with different individuals, but in all other cases there is the possibility that an individual's pollen may make contact with its own stigma. For this reason, most plants have genetic mechanisms to prevent fertilization from pollen grains that are too closely related to the stigma (self-incompatibility). The mechanisms of breeding systems occur at the molecular level through a biochemical reaction on the stigma that recognizes genetic differences in pollen grains. Depending on the species, individual plants can self-pollinate, individuals plants can cross-pollinate intraspecifically (between individuals of the same species), or individuals can cross-pollinate interspecifically (between individuals of different species) and hybridize. Orchids are known to have weak barriers to hybridization.

Self-incompatibility

Mating with individuals that are too closely related (i.e. with self) may result in inbreeding depression, so it is usually considered advantageous to cross-pollinate intraspecifically, in which case self-incompatibility is utilized. At least 27 families of monocots have genetic mechanisms to ensure self-incompatibility (SI). The most widespread form of self-incompatibility in monocots is gametophytic, meaning compatibility is determined by the genotype of the pollen grain. There are two described mechanisms of gametophytic self-incompatibility that have been shown to occur in four families of dicots (RNase and S-glycoprotein) but none have been found in monocots. However, there is evidence that orchids have an alternative undescribed mechanism of gametophytic self-incompatibility.

Homomorphic sporophytic self-incompatibility has not yet been discovered in monocots. In this form compatibility is determined by the genotype of the anther from which the pollen grain was created. Heteromorphic sporophytic self-incompatibility, a mechanism in heterostylous flowers, has been shown to occur in only one family of monocots, Pontederiaceae. Late-acting (ovarian) self-incompatibility has been described in Agavaceae, Iridaceae, and Amaryllidaceae.

Grasses have a mechanism of self-incompatibility unique to themselves; they employ two unlinked loci, S and Z. When the alleles at these loci are equivalent between a pollen grain and a stigma on which it lands then the pollen grain will be rejected.

Self-compatibility

Self-compatible (SC) pollination systems are less common than self-incompatibile cross-pollination systems in angiosperms. However, when the probability of cross-pollination is too low it can be advantageous to self-pollinate. Self-pollination is known to be favored in some orchids, rices, and Caulokaempferia coenobialis (Zingiberaceae).

Pollination Ecology

Pollination systems in monocots are just as diverse as in dicots. About two thirds of monocots evolved to be zoophilous (animal pollinated). Others are instead water-pollinated or wind-pollinated such as Cyperaceae, Juncaceae, Sparganiaceae, Typhaceae, and most notably Poaceae. These modes evolved to facilitate transfer of the pollen grain onto the stigma. Most zoophilous monocots are pollinator generalists with the most notable exception being the Orchids. Monocot pollen grains are monocolpate, meaning they have one groove; outer surfaces called exines are smooth.

Pollination Strategies

All monocots utilize either cross-pollination or self-pollination strategies, as do dicots, but the advantage of either strategy depends on ecological factors such as pollinator abundance and competition. These strategies either promote fertilization with self and suppress fertilization with others resulting in self-pollination, or they suppress fertilization with self and promote fertilization with others resulting in cross-pollination. Strategies also exist to suppress fertilization with other species as reproductive barriers.

Pollination strategies have the same function as breeding systems, however they occur at the ecological level or at the level of floral structure rather than at the molecular level on the stigma through genetic recognition.

Cross-pollination (Allogamy)

Self-pollination can be prevented by both physical and temporal mechanisms that have evolved in response to the interactions with pollen vectors; these mechanisms make cross-pollination easier to accomplish by lowering the chances of self-pollination. For example, dichogamy, which is the temporal differentiation in the ripening of sexual organs, is common in monocots with both protogynous and protoandrous flowers. Herkogamy, which is the spatial separation of sexual organs, is also present in many monocots.

Self-pollination (Autogamy and Geitonogamy)

Self-pollination can occur with or without the aid of animals. When animal-mediated, sexual organs will be positioned closer spatially and temporally, inverse to the strategies of dichogamy and herkogamy. However, when self-pollination is self-induced by the flower, some unique mechanisms have evolved. In Caulokaempferia coenobialis (Zingiberaceae), pollen is transported via a drop of oil that forms on the anther and slowly slides down to the stigma. In the orchid, Paphiopedilum parishii, anthers liquify and touch the stigma with the help of gravity rather than a pollinator. Another orchid, Holcoglossum amesianum, rotates its anther in circles to transfer pollen into its stigma cavity.

Apomixis (Agamospermy)

Apomixis is asexual reproduction through seeds and does not require pollination. It is distributed throughout the monocot clade in Poales, Asparagales, Liliales, Dioscoreales, and Alismatales. Thus apomixis may have evolved once in a basal ancestor and has since repeatedly become lost.

Pollination Syndromes

Pollination syndromes are floral adaptations in response to pollen vectors, such as the production of nectar.

The Allium flower has six stamens and six tepals.

Floral Morphology

Flower structure is more uniformly distributed within the monocots. Monocot flowers occur with parts having multiples of three; usually there are three stamen, three petals and three sepals (six tepals), and usually just one stigma. However stamens in twos can be found in Cypripedioideae while single fertile stamens can be found in Philydraceae, Zingiberaceae, and as gynostemium in Orchidacaea. Furthermore, flower structures that evolved to trap insects to accomplish pollination are found in many monocot genera. In relation to flower arrangement alone, plants with perfect flowers should be most likely to self-pollinate while dioecious plants should be most likely to cross-pollinate.

Animal Pollination

Zoophily, or animal pollination, is a method of pollination which utilizes animals as pollen vectors.

In order for pollen to affix to animal bodies, a tryphine coating is usually present in zoophilous pollen to achieve an adhesive pollen grain.

Visual attractants of monocot flowers mainly come from the coloration of tepals. However, when species with small green tepals are zoophilous other organs can evolve to be visually attractive such as having colored bracts (Araceae, Cyclanthaceae, and some

Arecaceae), otherwise attraction is based on scent only. The similar pigments used in monocot and dicot flower coloration have independently evolved.

Many monocots produce scent to attract pollinators but perhaps not as many as those that produce nectar.

A bee orchid, Ophrys, mimics a female bee as a false reward.

Most zoophilous monocots produce nectar as a reward and this nectar is alike to nectar of dicots. Carpellary septal nectaries are common and unique to monocots. Nonseptal nectaries are most often epithelial and positioned on the perigonal nectaries of tepals when occurring in monocots. Also, nectar can be produced in perigonal unicellular hairs, a trait only observed in monocots. Monocots do not have disc nectaries whereas in dicots they are widespread.

Like dicots, some zoophilous monocots do not produce nectar and instead offer pollen as the main reward. A few even offer other rewards: oils to bees, starchy tissue to beetles, sleeping holes to bees, and a perfume which Euglossini male bees will collect and present on their legs during mating displays. Deceptive flowers that do not offer actual rewards are much more widespread in monocots than dicots, with the most common perpetrator being the orchids. Orchids commonly provide empty nectar spurs. One genus, Ophrys, is known for its ability to mimic female bees to such a degree that it fools male bees into pseudocopulating with the "female" and thereby pollinating the flower.

Wind Pollination

Most wind-pollinated plants do not produce nectar, attractive scents, or petals because they are not adapted to pollination by animal vectors. Grasses are a large wind-pollinated group; their stigmas are often feathery to help catch pollen in the wind.

Water Pollination

Monocots account for nearly all hydrophilous or water-pollinated plants. These are monocots that are adapted to use water as a vector and constitute most of the aquatic plants. Depending on the species, pollen can either float on the surface and disperse by wind and water currents towards other surface-floating flowers, or pollen can drift underwater to flowers that are submerged. In the later scenario, pollen is without an exine and stigmas are forked.

The Flowering Bamboo Phenomenon

A few species of bamboos can grow for more than 120 years without flowering. Then at once flowering can simultaneously occur in groves across the world, termed gregarious or mast flowering. This is possible because the trigger to flower is genetically determined and because multiple forests can develop from the clones of one individual. The cause of the trigger is still unknown and unpredictable. During anthesis, or flowering, pollination is wind-mediated but bee pollination has been observed in at least 6 species. When pollination is zoophilous flowers can be fragrant and attract large numbers of pollinator-collecting bees to congregate around the inflorescence and take advantage of this new and abundant source of pollen. After anthesis massive die-offs of all sister groves occur within three years of each other and can have devastating effects.

AUTOGAMY

Autogamy, or self-fertilization, refers to the fusion of two gametes that come from one individual. Autogamy is predominantly observed in the form of self-pollination, a reproductive mechanism employed by many flowering plants. However, species of protists have also been observed using autogamy as a means of reproduction. Flowering plants engage in autogamy regularly, while the protists that engage in autogamy only do so in stressful environments.

Protists

Paramecium Aurelia

Paramecium aurelia is the most commonly studied protozoan for autogamy. Similar to other unicellular organisms, Paramecium aurelia typically reproduce asexually via binary fission or sexually via cross-fertilization. However, studies have shown that when put under nutritional stress, Paramecium aurelia will undergo meiosis and subsequent fusion of gametic-like nuclei. This process, defined as hemixis, a chromosomal rearrangement process, takes place in a number of steps. First, the two micronuclei of Paramecium aurelia enlarge and divide two times to form eight nuclei. Some of these daughter nuclei will continue to divide to create potential future gametic nuclei. Of these potential gametic nuclei, one will divide two more times. Of the four daughter nuclei arising from this step, two of them become anlagen, or cells that will form part of the new organism. The other two daughter nuclei become the gametic micronuclei that will undergo autogamous self-fertilization. These nuclear divisions are observed mainly when the Paramecium aurelia is put under nutritional stress. Research shows that Paramecium aurelia undergo autogamy synchronously with other individuals of the same species.

Clonal Aging and Rejuvenation

In Paramecium tetraurelia, vitality declines over the course of successive asexual cell divisions by binary fission. Clonal aging is associated with a dramatic increase in DNA damage. When paramecia that have experienced clonal aging undergo meiosis, either during conjugation or automixis, the old macronucleus disintegrates and a new macronucleus is formed by replication of the micronuclear DNA that had just experienced meiosis followed by syngamy. These paramecia are rejuvenated in the sense of having a restored clonal lifespan. Thus it appears that clonal aging is due in large part to the progressive accumulation of DNA damage, and that rejuvenation is due to repair of DNA damage during meiosis that occurs in the micronucleus during conjugation or automixis and reestablishment of the macronucleus by replication of the newly repaired micronuclear DNA.

Tetrahymena Rostrata

Similar to Paramecium aurelia, the parasitic ciliate Tetrahymena rostrata has also been shown to engage in meiosis, autogamy and development of new macronuclei when placed under nutritional stress. Due to the degeneration and remodeling of genetic information that occurs in autogamy, genetic variability arises and possibly increases an offspring's chances of survival in stressful environments.

Allogromia Laticollaris

Allogromia laticollaris is perhaps the best-studied foraminiferan amoeboid for autogamy. Allogromia laticollaris can alternate between sexual reproduction via cross-fertilization and asexual reproduction via binary fission. The details of the life cycle of Allogromia laticollaris are unknown, but similar to Paramecium aurelia, Allogromia laticollaris is also shown to sometimes defer to autogamous behavior when placed in nutritional stress. As seen in Paramecium, there is some nuclear dimorphism observed in Allogromia laticollaris. There are often observations of macronuclei and chromosomal fragments coexisting in Allogromia laticollaris. This is indicative of nuclear and chromosomal degeneration, a process similar to the subdivisions observed in Paramecium aurelia. Multiple generations of haploid Allogromia laticollaris individuals can exist before autogamy actually takes place. The autogamous behavior in Allogromia laticollaris has the added consequence of giving rise to daughter cells that are substantially smaller than those rising from binary fission. It is hypothesized that this is a survival mechanism employed when the cell is in stressful environments, and thus not able to allocate all resources to creating offspring. If a cell was under nutritional stress and not able to function regularly, there would be a strong possibility of its offspring's fitness being sub-par.

Self-pollination in Flowering Plants

Self-pollination is an example of autogamy that occurs in flowering plants. Self-pollination occurs when the sperm in the pollen from the stamen of a plant goes to the carpels of that same plant and fertilizes the egg cell present. Self-pollination can either be done completely autogamously or geitonogamously. In the former, the egg and sperm cells that united came from the same flower. In the latter, the sperm and egg cells can come from a different flower on the same plant. While the latter method does blur the lines between autogamous self-fertilization and normal sexual reproduction, it is still considered autogamous self-fertilization.

Self-pollination can lead to inbreeding depression due to expression of deleterious recessive mutations. Meiosis followed by self-pollination results in little genetic variation, raising the question of how meiosis in self-pollinating plants is adaptively maintained over an extended period in preference to a less complicated and less costly asexual ameiotic process for producing progeny. For instance, Arabidopsis thaliana is a predominantly self-pollinating plant that has an outcrossing rate in the wild estimated at less than 0.3%, and self-pollination appears to have evolved roughly a million years ago or more. An adaptive benefit of meiosis that may explain its long-term maintenance in self-pollinating plants is efficient recombinational repair of DNA damage.

Advantages of Autogamy

There are several advantages for the self-fertilization observed in flowering plants and protists. In flowering plants, it is important for some plants not to be dependent on pollinating agents that other plants rely on for fertilization. This is unusual, however, considering that many plant species have evolved to become incompatible with their own gametes. While these species would not be well served by having autogamous self-fertilization as a reproductive mechanism, other species, which do not have self-incompatibility, would benefit from autogamy. Protists have the advantage of diversifying their modes of reproduction. This is useful for a multitude of reasons. First, if there is an unfavorable change in the environment that puts the ability to deliver offspring at risk, then it is advantageous for an organism to have autogamy at its disposal. In other organisms, it is seen that genetic diversity arising from sexual reproduction is maintained by changes in the environment that favor certain genotypes over others. Aside from extreme circumstances, it is possible that this form of reproduction gives rise to a genotype in the offspring that will increase fitness in the environment. This is due to the nature of the genetic degeneration and remodeling intrinsic to autogamy in unicellular organisms. Thus, autogamous behavior may become advantageous to have if an individual wanted to ensure offspring viability and survival. This advantage also applies to flowering plants. However, it is important to note that this change has not shown to produce a progeny with more fitness in unicellular organisms. It is possible that the nutrition deprived state of the parent cells before autogamy created a barrier for producing offspring that could thrive in those same stressful environments.

Disadvantages of Autogamy

In flowering plants, autogamy has the disadvantage of producing low genetic diversity in the species that use it as the predominant mode of reproduction. This leaves those species particularly susceptible to pathogens and viruses that can harm it. In addition, the foraminiferans that use autogamy have shown to produce substantially smaller progeny as a result. This indicates that since it is generally an emergency survival mechanism for unicellular species, the mechanism does not have the nutritional resources that would be provided by the organism if it were undergoing binary fission.

References

- Plant-reproductive-system, science: britannica.com, Retrieved 15 May, 2020
- Asexual-reproduction, boundless-biology-chapter: courses.lumenlearning.com, Retrieved 29 June, 2020
- Eckert CG (2000). "Contributions of Autogamy and Geitonogamy to Self-Fertilization in a Mass-Flowering, Clonal Plant". Ecology. 81 (2): 532. doi:10.2307/177446
- Micro-propagation-technique-factors-applications-and-disadvantages, biotechnology-clonal-propagation-10732: biologydiscussion.com, Retrieved 03 April, 2020
- Sexual-reproduction-in-plants, wmopen-biology2-chapter: courses.lumenlearning.com, Retrieved 23 February, 2020
- Barrett, S.C.H. (2002). "The evolution of plant sexual diversity" (PDF). Nature Reviews Genetics. 3 (4): 274–284. doi:10.1038/nrg776. PMID 11967552

Plant Breeding

CHAPTER 5

The science of altering traits of plants to create desired genotypes and phenotypes for improving the genetic potential of plants is termed as plant breeding. It involves controlled pollination and genetic engineering followed by artificial selection of progeny. This chapter delves into plant breeding for an in-depth understanding of the subject.

Plant breeding is the use of natural and artificial selection to produce heritable variations and novel combinations of alleles in plants and to identify plants with novel and useful properties. The goals of plant breeding are to produce crop varieties that boast unique and superior traits for a variety of agricultural applications. The most frequently addressed traits are those related to biotic and abiotic stress tolerance, grain or biomass yield, end-use quality characteristics such as taste or the concentrations of specific biological molecules (proteins, sugars, lipids, vitamins, fibers) and ease of processing (harvesting, milling, baking, malting, blending, etc).

The specific goal of a plant breeding project is highly dependent upon the market for which the product is intended. In wheat, for example, varieties bred to produce high levels of gluten protein are used to produce noodles and breads, whereas varieties specifically selected for low protein grains are used in the production of pastry flours. The varieties of grapes developed for various applications such as juices, red and white wines, and jams, vary considerably in chemical composition.

Plant breeders employ a variety of techniques to improve the genetic composition of the crop and a successful strategy is dependent on the physical, physiological and hereditary characteristics of the plant. Different breeding approaches are required for self pollinating, cross-pollinating and clonally-propagated plants. The methods used by plant breeders have developed along with the advancement of human civilization and have expanded to incorporate humanity's increased knowledge of genetics.

Pre-mendelian Plant Breeding

Plant breeding has been practiced for over ten millennia, and arose independently in several centers, including South and Central America, Asia, the Fertile Crescent and New Guinea. While early farmers recognized that, at least for inbred crops, "like begat like," the genetic aspects of plant breeding were enigmatic until the relatively recent work of Mendel. As such, early plant breeders practiced a form of mass selection, in which plants of superior phenotype were selected and seed of these plants was replanted during the next season.

Prior to the 18th century there is little evidence of the intentional, controlled crosses between plant varieties which typify modern plant breeding. Thus, the lines developed by early plant breeders are what we would now term landraces, locally adapted lines shaped largely through either natural selection or farmer-driven selection in a particular region; these are colloquially known as folk varieties.

The first implementation of controlled crosses is attributed to the late 18th century German botanist Joseph Kolreuter. However, his study used the ornamental plant Dianthus herbatus and thus was more botanical in nature rather than aimed at the improvement of a crop population. The use of controlled pollination to make new varieties was employed with increasingly greater precision during the 19th century, particularly in maize which received wide attention from early plant breeders due to its many visible phenotypes. However, while the physical principles of plant reproduction were better elucidated by botanists during the late 18th and early 19th century, the underlying mechanisms of inheritance were still enigmatic.

Theories regarding heredity abounded in the 19th century among both zoologists and botanists. In 1856 Louis de Vilmorin published his studies on heredity in beetroot and reported that progeny from two seemingly similar plants can exhibit emerging traits. De Vilmorin was thus among the first to recognize the distinction between genotype and phenotype although he did not yet formally identify these features as such.

The work of these pre-Mendelian plant biologists represented the early synthesis of botany and heredity into the emerging sub-discipline of plant breeding. However, it was not until the work of Mendel that modern techniques of quantitatively manipulating heredity of plant populations became possible.

Post-Mendelian Plant Breeding

Mendel's 1865 paper outlining and quantifying the principles of inheritance provided the basis for all modern genetics, but was sadly undiscovered until the beginning of the 20th century. As human understanding of inheritance and the gene action rapidly deepened, the application of these principles to crop production was pursued with equal alacrity.

A mere three years after the rediscovery of Mendel in 1900 by De Vries, Correns and von Tschermark, the Danish botanist Wilhelm Johannsen developed the pure-line breeding theory that is employed to generate true-breeding lines through repeated self-pollination. Johannsen also defined the role of environment in the inheritance of quantitative traits, a concept that is critical to the proper execution of field experiments in plant breeding. Marcus Rhoades, in 1933, incorporated the use of cytoplasmic male sterility genes into maize breeding; this permits the controlled crossing of plants without the need for a labor-intensive removal of male reproductive structures. The establishment of the Haber-Bosch process in 1910 to artificially produce ammonia provided plant-breeders with the new goal of adapting varieties to take advantage of abundant amounts of nitrogen fertilizer.

In the latter half of the 20th century, plant breeders incorporated techniques aimed at increasing diversity and expediting the breeding process. Following the works of Thomas Hunt Morgan and Hermann Muller in the 1910's and 1920's, plant breeders saw the value of physical and chemical mutagenesis for the production of new phenotypes. The application of statistics and population genetics to plant breeding was accomplished through the work of scientists such as Nikolai Vavilov and J.R. Harlan. The elucidation of DNA as the genetic material by Watson, Crick and Wilkins, and the later deduction of the genetic code further empowered the efforts of plant breeders through increased comprehension of gene action.

Transfer of genetic material between disparate species was aided by the use of tissue culture methods and later through transgenic technologies. Initiatives and developments in genomics made possible more rapid identification and targeting of gene sequences conferring desirable qualities. These methods are all discussed in detail in later sections.

Perhaps the most significant advance in plant breeding in the 20th century, and the most defining time for modern plant breeding was the so-called Green Revolution, the dramatic increase in crop yields that occurred between 1945 and 1985, credited in large part to the work of plant breeder Norman Borlaug. While numerous factors, both scientific and political, contributed to the progress of the Green Revolution, the greatest impact came from the development of high-yielding varieties in several of the world's major crops. This was accomplished through alterations to both plant genotypes and farming practices. Yields were dramatically increased in maize, wheat and rice by breeding for both increased nitrogen assimilation and reduced height. These so-called semi-dwarf varieties are deficient in components of the gibberilic acid pathways and thus have limited apical dominance. The net result is the investment of energy into grains without lodging (i.e. falling over before harvest).

The Norin 10 wheat cultivar, produced by plant breeder Orville Vogel, formed the basis for increasing yields in the world's most abundant crop, and the Filipino high-yielding rice cultivar IR8 also had a significant impact on world food production. The increased application of fertilizers, pesticides, and the rise of irrigation and mechanization also contributed to increased yields and improved market efficiency. It is accurate to say that modern plant breeding goals and methods have been set within the context established by the Green Revolution; this includes meeting the challenges and problems which the Green Revolution failed to solve (or in some cases, exacerbated).

Genetic Basis of Plant Breeding

Mendel was able to successfully quantify the inheritance of traits primarily because he chose an easily manipulated crop for his experiments: the green garden pea. His investigation used self-pollinating (autogamous) plants and he assessed simply inherited, qualitative traits in a controlled environment. Plant breeders seldom have such luxury, as most agronomically significant traits are determined by complex gene interactions. Furthermore, many important crop species do not appreciably selfpollinate, instead requiring cross-pollination (allogamous), while others generally undergo vegetative propagation. Wheat, rice, soybean, sorghum and millets are examples of self-pollinating crop plants; corn, sugarcane and sugar beet are crosspollinating; coconut is cross-pollinating or self-pollinating; potato, sweet potato, cassava, grape, banana and orange have a vegetative mode of reproduction.

Mendel's work helped establish the modern terms "homozygous", indicating an organism with two identical alleles for a given gene, and "heterozygous", indicating an organism with two different alleles for a gene. The term "true-breeding" is often applied to homozygotes, because when self-pollinated (or inbred in the case of cross-pollinating species), these individuals only produce more homozygotes. Self-pollinating heterozygotes, on the other hand, produce 50% heterozygotes and 25% homozygotes of each class.

The plant breeder must also consider the environment in which the plant will be cultivated; for example a variety which boasts high quality under irrigation may well be a poor performer in an arid region. While these factors may make Mendelian inheritance seem inadequate to describe plant breeding efforts, these principles, though in a greatly complicated form, comprise the basis of a crop breeding project. Thus, the fundamental principles of plant breeding can be explained using Mendelian analysis, if we lay aside greater complexities for the moment.

The standard methodology of plant breeding is to create a population of segregating alleles in order to select superior combinations from a diverse population, thus heterozygosis is desired in early generations. However, true-breeding lines, homozygous at all loci, are desired as both the starting materials and end-products of this process, at least when breeding self-pollinating species.

A standard plant breeding project begins with a cross of two parental plant varieties to create an F_1 hybrid population, heterozygous at many loci, which will segregate in subsequent generations. This is generally accomplished through emasculation, removal of the pollen-bearing structures, of one plant to ensure that the two plants cross-fertilize. This is generally done twice, with each parental genotype type serving as the pollen donor in the respective crosses. This is termed a reciprocal cross and is important to measure the effects of cytoplasmic inheritance (mitochondria and chloroplasts are always inherited from the female parents).

In maize, de-tasseling, the removal of the pollen-bearing tassels of the plant, is labor intensive, and the cunning implementation of cytoplasmic male sterility (CMS) genes into the breeding process significantly reduces the effort needed to insure cross pollination. This method is currently used when genetically modified and non-GM crops are grown together, to prevent transfer of transgenes. For self-pollinating species, the parental plants of a cross should be true-breeding, so that an isogenic (genetically identical) F_1 population is produced. If the parental lines are themselves still segregating, the F_1 will not be uniform and some alleles may not be passed on to any F_2 individuals, thus sacrificing diversity. In subsequent generations, the breeder will select segregating lines that demonstrate desired qualities, refining their selection with more stringent criteria in successive rounds of selection.

Genetic recombination, or crossing over, between chromosomes in the early generations of a project allows new combinations of alleles by breaking up linkage blocks, clusters of alleles at closely-linked loci on a chromosome. In general, crosses between genetically distinct lines with differing, desirable characteristics are most successful because segregation produces a population with high variation. Ultimately, for selfpollinated crops, most breeders release true-breeding lines, also called pure lines. Breeders of wheat or rice, for instance, would not release lines for cultivation that retain heterozygosis, as subsequent generations may lose key characteristics through segregation. Fortunately, this task is easy with self-pollinating species, as the heterozygosis of a population is halved in each generation of self-fertilization.

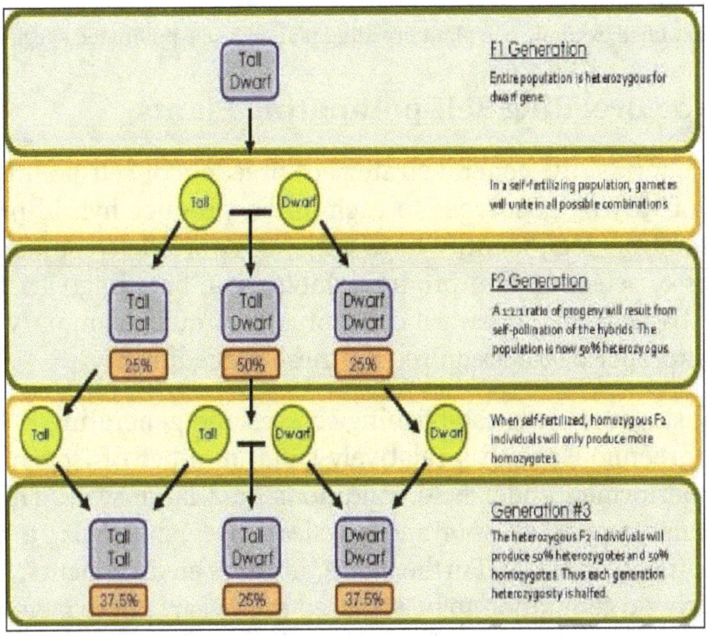

Self-pollination of hybrid plants leads to a halving of heterozygosis in each generation. After six generations, approximately 97% will be homozygous.

Thus in order to establish a true-breeding line, the breeder of a self-pollinated crop merely has to engage in repeated cycles of self-fertilization and selection.

This loss of heterozygosis has prompted the designation of self-pollinating lines in the F5 or later generations as recombinant inbred lines, because they are the product of recombination in a segregating population, but are likely homozygous at many loci due to inbreeding.

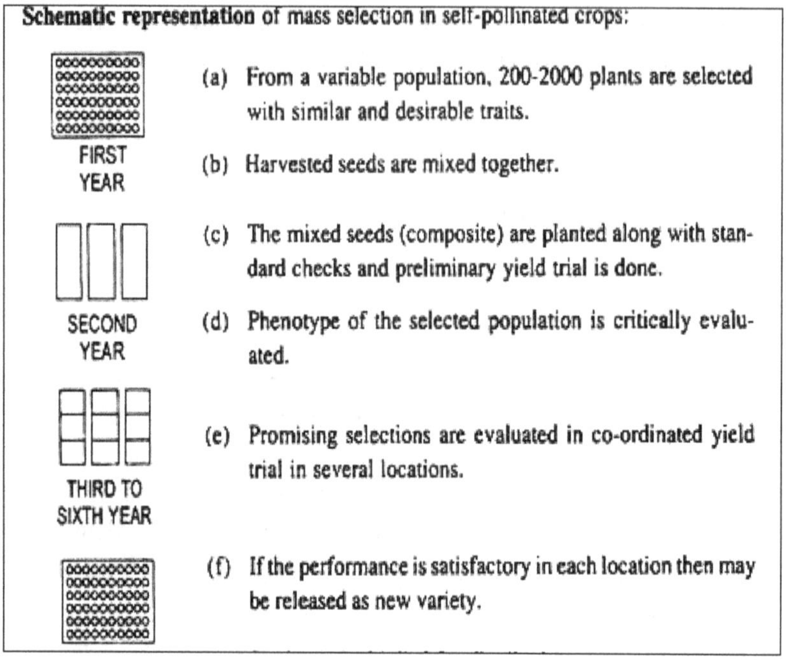

Simplified example of a plant breeding plan in a self-pollinated species.

Techniques for Breeding Self-pollinating Plants

In the most basic terms, the general strategy of breeders of self-pollinated crops is to cross two lines of varying genotypes to each other, produce hybrid progeny that are segregating at multiple loci, identify lines with superior allele combinations through successive cycles of selection and produce stable, true-breeding varieties from these lines. While relatively straightforward conceptually, a quick summary fails to convey the physical and temporal scale required for these procedures.

While the initial stages such as establishing crosses and generating F_1 hybrids, can be performed in a greenhouse with a relatively small number of isogenic plants, latter stages must be performed under field conditions on a large scale. This obviously requires a significant amount of labor and physical space, particular to implement adequate replication and controls. Furthermore, in most environments, field conditions allow for only a single generation to be assessed each year. These factors place considerable constraints on the efficiency of plant breeding projects.

As such, modern plant breeding techniques are largely focused on two goals: 1) Increasing the efficiency of the selection process and production of true-breeding lines; and 2) Acquisition or de novo generation of additional genetic diversity in plant populations. Methods used to achieve the former goal will be considered first.

Benefits of Plant Breeding

Experienced plant breeders have made great scientific strides to improve the crops you grow. Innovation in plant breeding delivers higher yielding varieties with better agronomic traits, such as disease resistance and stress tolerance, to benefit your farm and bottom line.

- Plant breeding brought wheat to the West in 1906 with the variety Marquis paving the way for even more improved varieties.
- Higher yields with hybrid corn and many vegetable crops. Over the past 60 years, the efforts of corn breeders have been rewarded by a 400% increase in corn yields.
- New cropping options with canola. Canola didn't exist prior to 1974; it now contributes between $1.5 and $2 billion to Canada's agricultural economy.
- Improved time management and on-farm erosion control with herbicide tolerant crops.
- New high-value markets in Japan with specialty soybeans.
- Global competitiveness through better yields and high quality.
- Today, plant breeders are working on even more problems that can increase your net farm income:
    - Fusarium head blight control for wheat growers.
    - Canola seeds with increased vigour could reduce crop establishment risk.
    - Drought resistance is being developed in canola, corn and soybeans.
    - Increased markets with specialty crops to supply industrial products like hydraulic oils, bio-diesel, and plastics – create biodegradable products from renewable resources.
    - Value-added opportunities for crops with increased nutritional content or healthier qualities.

METHODS OF PLANT BREEDING

Conventional Methods

Plant breeding is defined as identifying and selecting desirable traits in plants and combining these into one individual plant. Since 1900, Mendel's laws of genetics provided the scientific basis for plant breeding. As all traits of a plant are controlled by genes located on chromosomes, conventional plant breeding can be considered as the manipulation of the combination of chromosomes. In general, there are three main procedures to manipulate plant chromosome combination. First, plants of a given population which show desired traits can be selected and used for further breeding and cultivation, a process called (pure line) selection. Second, desired traits found in different plant lines can be combined together to obtain plants which exhibit both traits simultaneously, a method termed hybridization. Heterosis, a phenomenon of increased vigor, is obtained by hybridization of inbred lines. Third, polyploidy (increased number of chromosome sets) can contribute to crop improvement. Finally, new genetic variability can be introduced through spontaneous or artificially induced mutations.

- Selection: Selection is the most ancient and basic procedure in plant breeding. It generally involves three distinct steps. First, a large number of selections are made from the genetically variable original population. Second, progeny rows are grown from the individual plant selections for observational purposes. After obvious elimination, the selections are grown over several years to permit observations of performance under different environmental conditions for making further eliminations. Finally, the selected and inbred lines are compared to existing commercial varieties in their yielding performance and other aspects of agronomic importance.

- Hybridization: The most frequently employed plant breeding technique is hybridization. The aim of hybridization is to bring together desired traits found in different plant lines into one plant line via cross- pollination. The first step is to generate homozygous inbred lines. This is normally done by using self-pollinating plants where pollen from male flowers pollinates female flowers from the same plants. Once a pure line is generated, it is outcrossed, i. e. combined with another inbred line. Then the resulting progeny is selected for combination of the desired traits. If a trait from a wild relative of a crop species, e.g. resistance against a disease, is to be brought into the genome of the crop, a large quantity of undesired traits (like low yield, bad taste, low nutritional value) are transferred to the crop as well. These unfavorable traits must be removed by time-consuming back-crossing, i. e. repeated crossing with the crop parent. There are two types of hybrid plants: interspecific and intergeneric hybrids. Beyond this biological boundary, hybridization cannot be accomplished due to sexual incompatibility, which limits the possibilities of introducing desired traits into crop Plants.

- Heterosis is an effect which is achieved by crossing highly inbred lines of crop plants. Inbreeding of most crops leads to a strong reduction of vigor and size in the first generations. After six or seven generations, no further reduction in vigor or size is found. When such highly inbred plants are crossed with other inbred varieties, very vigorous, large sized, large-fruited plants may result. The term "heterosis" is used to describe the phenomenon of hybrid vigor. The most notable and successful hybrid plant ever produced is the hybrid maize. By 1919, the first commercial hybrid maize was available in the United States. Two decades later, nearly all maize was hybrid, as it is today, although the farmers must buy new hybrid seed every year, because the heterosis effect is lost in the first generation after hybridization of the inbred parental lines.

- Polyploidy: Most plants are diploid. Plants with three or more complete sets of chromosomes are common and are referred to as polyploids. The increase of chromosomes sets per cell can be artificially induced by applying the chemical colchicine, which leads to a doubling of the chromosome number. Generally, the main effect of polyploidy is increase in size and genetic variability. On the other hand, polyploid plants often have a lower fertility and grow more slowly.
- Induced Mutation: Instead of relying only on the introduction of genetic variability from the wild species gene pool or from other cultivars, an alternative is the introduction of mutations induced by chemicals or radiation. The mutants obtained are tested and further selected for desired traits. The site of the mutation cannot be controlled when chemicals or radiation are used as agents of mutagenesis. Because the great majority of mutants carry undesirable traits, this method has not been widely used in breeding programs.

Success in using conventional plant breeding principles and agricultural techniques reached its peak when high-yielding wheat and rice lines were cultivated in the 1960s. The doubling and tripling of productivity of these important crops in Asia signaled a agricultural revolution in the developing countries. This breakthrough in food production was termed the "Green Revolution" to describe the social, economic, and nutritional impact of the new high-yielding wheat and rice strains. Norman Borlaug was awarded the Noble Price in 1970 for his contribution in breeding new high-yielding strains of cereals. However, these strains were highly dependent on fertilizers, irrigation and agrochemicals and required energy-intensive investments. This led to degradation and loss of soils and other severe environmental problems all over the world. For these reasons, the "Green Revolution" has been both praised and damned.

Biotechnological Methods

Biotechnology is the discipline which deals with the use of living organisms or their products. In this wide sense, also traditional agriculture may be seen as a form of biotechnology. The European Federation of Biotechnology defines biotechnology as "the integrated use of biochemistry, microbiology and engineering sciences in order to achieve technological (industrial) application of the capability of microorganisms, cultured tissue cells and parts thereof". In recent years, biotechnology has developed rapidly as a practical means for accelerating success in plant breeding and improving economically important crops.

- In Vitro Cultivation of Plant Cells and Regeneration of Plants from Cultured Cells Certain isolated somatic plant cells can be cultured in vitro (in the test tube) and are capable of proliferation and organization into tissues and eventually into complete plants. The process of regenerating whole plants out of plant cells is called in vitro regeneration. The three factors affecting plant regeneration are genotype, explant source, and culture conditions, including culture medium and environment. Different mixtures of plant hormones and other compounds in varying concentrations are used to achieve regeneration of plants from cultured cells and tissues. As the plant hormonal mechanisms are not yet understood completely, the development of in vitro cultivation and regeneration systems is still largely based on empirically testing variations of the three above mentioned factors.

Vitro Selection and Somaclonal Variation: Plants regenerated from cell cultures may exhibit phenotypes differing from their parent plants, sometimes at quite high frequencies. If these are heritable and affecting desirable agronomic traits, such "somaclonal variation" can be incorporated into breeding programs. However, finding of specific valuable traits by this method is largely left to chance and hence inefficient. Rather than relying on this undirected process, selection in vitro targets specific traits by subjecting large populations of cultured cells to the action of a selective agent in the Petri dish. For purposes of disease resistance, this selection can be provided by pathogens, or isolated pathotoxins that are known to have a role in pathogenesis. The selection will only allow those cells to survive and proliferate that are resistant to the challenge. Selection of cells also plays an important role in genetic engineering, where special marker genes are used to select for transgenic cells.

Somatic Hybrid Plants: Somatic hybrid plants are plants derived from the fusion of somatic cells. Cell fusion was developed after the successful culture of a large number of plant cells stripped of their cell walls. The resulting cells without walls are referred to as protoplasts. Since also protoplasts from phylogenetically unrelated species can be fused, attempts have been made to overcome sexual incompatibility using protoplast fusion. In most cases, these attempts failed because growth and division of the fused cells did not take place when only distantly related cells were fused. Successful fusions between sexually incompatible petunia species and between potatoes and tomatoes did not lead to economically interesting products, but important contributions to the understanding of cell wall regeneration and other mechanisms were achieved.

Genetic Engineering

Genetic engineering is a term used for the directed manipulation of genes, i. e. the transfer of genes between organisms or changes in the sequence of a gene. Closely related to this field are methods which use genes or specific sequences for the identification of traits and other analytical purposes. In plant breeding, the most important and already widely used method of this kind is Restriction Fragment Length Polymorphism (RFLP).

- Restriction Fragment Length Polymorphism (RFLP): The techniques of traditional breeding are very time-consuming. By making crosses, also a large number of undesired genes is introduced into the genome of the plant. The undesired genes have to be "sorted out" by back-crossing. The use of Restriction Fragment Length Polymorphism greatly facilitates conventional plant breeding, because one can progress through a breeding program much faster, with smaller populations and without relying entirely on testing for the desired phenotype.

RFLP makes use of restriction endonucleases. These are enzymes which recognize and cut specific nucleotide sequences in DNA. For example, the sequence GAATTC is cut by the endonuclease EcoRl. After treatment of a plant genome which restriction endonucleases, the plant DNA is cut into pieces of different length, depending on the number of recognition sites on the DNA. These fragments can be separated according to their size by using gel electrophoresis and are made visible as bands on the gel by hybridizing the plant DNA fragments with radiolabeled or fluorescent DNA probes. As two genomes are not identical even within a given species due to mutations, the number of restriction sites and therefore the length and numbers of DNA fragments differ, resulting in a different banding pattern on the electrophoresis gel. This variability has been termed restriction fragment length polymorphism (RFLP). The closer two organisms are related, the more the pattern of bands overlap. If a restriction site lies close to or even within an important gene, the existence of a particular band correlates with the particular trait of a plant, e.g. disease resistance. By looking at the banding pattern, breeders can identify individuals which have inherited resistance genes, and resistant plants can be selected for further breeding. The use of this technique will not only accelerate progress in plant breeding considerably, but will also facilitate the identification of resistance genes, thereby opening new possibilities in plant breeding.

Gene Transfer

In conventional breeding, the pool of available genes and the traits they code for is limited due to sexual incompatibility to other lines of the crop in question and to their wild relatives. This restriction can be overcome by using the methods of genetic engineering, which in principle allow introducing valuable traits coded for by specific genes of any organism (other plants, bacteria, fungi, animals, viruses) into the genome of any plant. The first gene transfer experiments with plants took place in the early 1980s. Normally, transgenes are inserted into the nuclear genome of a plant cell. Recently it has become possible to introduce genes into the genome of chloroplasts and other plastids (small organelles of plant cells which possess a separate genome).

Transgenic plants have been obtained using Agrobacterium-mediated DNA-transfer and direct DNA-transfer, the latter including methods such as particle bombardment, electroporation and polyethylenglycol permeabilisation. The majority of plants have been transformed using Agrobacterium mediated transformation.

- Agrobacterium-mediated Gene Transfer: The Agrobacterium-mediated technique involves the natural gene transfer system resident in the bacterial plant pathogens of the genus Agrobacterium. In nature, Agrobacterium tumefaciens and Agrobacterium rhizogenes are the causative agents of the crown gall and the hairy root diseases, respectively. The utility of Agrobacterium as a gene transfer system was first recognized when it was demonstrated that these plant diseases were actually produced as a result of the transfer and integration of genes from the bacteria into the genome of the plant. Both Agrobacterium species carry a large plasmid (small circular DNA molecule) called Ti in A. tumefaciens and Ri in A. rhizogenes. A segment of this plasmid, designated T-(for transfer) DNA, is transmitted by this organism into individual plant cells, usually within wounded tissue. The T-DNA segment penetrates the plant cell nucleus and integrates randomly into the genome where it is stably incorporated and inherited like any other plant gene in a predictable, dominant Mendelian fashion. Expression of the natural genes on the T-DNA results in the synthesis of gene products that direct the observed morphological changes, i. e. tumor or hairy root formation.

In genetic engineering, the tumor inducing genes within the T-DNA which cause the plant disease are removed and replaced by foreign genes. These genes are then stably integrated into the genome of the plant after infection with the altered strain of Agrobacterium, just like the natural T-DNA. Because all tumor-inducing genes are removed, the gene transfer does not induce any disease symptoms.This reliable method of gene transfer is well suited for plants which are susceptible to infection by Agrobacterium. Unfortunately, many species, especially economically important legumes and monocotyledons such as cereals, do not respond positively to Agrobacterium-mediated transformation. For these plants, the following methods of direct DNA uptake must be applied.

- Particle Bombardment: This method, also referred to as biolistic transformation involves coating biologically active DNA onto small tungsten or gold particles and accelerating them into plant tissue at high velocity. The particles penetrate the plant cell wall and lodge themselves within the cell where the DNA is liberated resulting in transformation of the individual plant cell in an explant. This technique is generally less efficient than Agrobacterium-mediated transformation, but has nevertheless been particularly useful in several plant species, most notably in cereal crops. The introduction of DNA into organized, morphogenic tissues such as seeds, embryos or meristems has enabled the successful transformation and regeneration of rice, wheat, soybean and maize, thus demonstrating the enormous potential of this method.
- Electroporation and Direct DNA Entry into Protoplasts: Electroporation is a process whereby very short pulses of electricity are used to reversibly permeabilize lipid bilayers of plant cell membranes. The electrical discharge enables the diffusion of macromolecules such as DNA through an otherwise impermable plasma membrane. Because the plant cell wall will not allow the efficient diffusion of many transgene constructs, protoplasts (cells without cell walls) must be prepared. This requirement presents a major obstacle for many applications as

protocols making possible the regeneration of protoplasts into complete plants do not exist for many species.

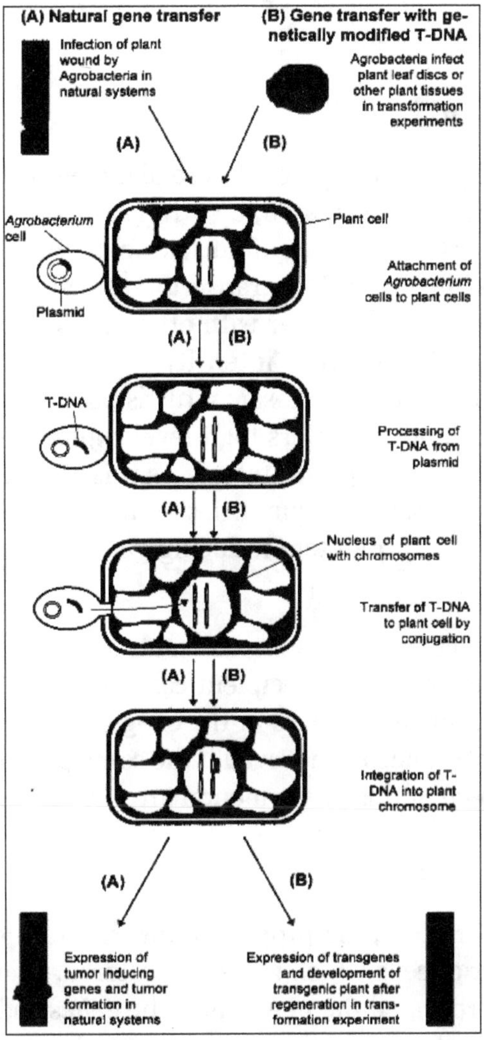

Gene transfer from Agrobacterium tumefaciens to a plant cell. In nature (route A), the transfer DNA (T-DNA) contains tumor inncluding genes which lead to crown gall disease. In plant genetic engineering (route B), the tumor including genes are removed and replaced by other genes, e.g. genes which confer insect resistance. In nature as well as in genetic engineering, the agrobacteria attach to a plant cell. Then, the T-DNA is cut out of a plasmid (a small circular DNA-molecule) and is transfered to the plant cell. The T-DNA migrates to the nucleus of the plant cell and becomes incorporated into a plant chromosome. In nature (route A), the tumor inducing genes cause uncontrolled (tumor-like) growth at the site of infection after expression of the T-DNA. In genetic engineering (route B, tumor inducing genes removed by other genes), transgenic cells are selected and regenerated to whole plants. All cells of the regenerated plant now contain the transgene.

DNA uptake by plant protoplasts can also be stimulated by phosphate or calcium/ polyethylene glycol (PEG) coprecipitation. However, these methods all suffer from the drawback that they use protoplasts as the recipient host which often cannot be regenerated into whole plants.

Transgene Expression

In most cases, the introduction of a gene into the plant genome will only have an effect on the plant if the transgene is expressed, i. e. transcribed into mRNA and translated into a protein. A promoter is a sequence of nucleic acids where the RNA polymerase (a complex enzyme synthesizing the mRNA transcript) attaches to the DNA template. The nature of the promoter defines (together with other expression-regulating elements), under which conditions and with which intensity a gene will be transcribed. The promoter of the 35S gene of cauliflower mosaic virus is used very frequently in plant genetic engineering. This promoter confers high-level expression of exogenous genes in most cell types from virtually all species tested. As it is often advantageous to express a transgene only in certain tissues or quantities or at certain times, a number of other promoters are available, e.g. promoters inducing gene expression after wounding or during fruit ripening only.

Methods of gene transfer currently employed result in the random integration of foreign DNA throughout the genome of recipient cells. The site of insertion may have a strong influence on the expression levels of the exogenous gene, resulting in different expression levels of an introduced gene, even if the same promoter/gene construct was used. The exact mechanism of this phenomenon are not yet fully understood.

Selection and Plant Regeneration

In a transformation experiment, the proportion of transformed cells is usually small compared to the number of cells which remain unaltered. In order to select only cells which have actually incorporated the new genes, the genes coding for the desired trait are fused to a gene which allows selection of transformed cells, so-called marker genes. The expression of the marker gene enables the transgenic cells to grow in presence of a selective agent, usually an antibiotic or a herbicide, while cells without the marker gene die. One of the most commonly used marker genes is the bacterial aminoglycoside-3' phosphotransferase gene (APH(3')II), also referred to as neomycin phosphotransferase 11 (NPTII). This gene codes for an enzyme which inactivates the antibiotics kanamycin, neomycin and G418 through phosphorylation. In addition to NPTII, a number of other antibiotic resistance genes have been used as selective markers, e.g. hygromycin phosphotransferase gene conferring resistance to hygromycin.

Another group of selective markers are herbicide tolerance genes. Herbicide tolerance has been obtained through the incorporation and expression of a gene which either detoxifies the herbicide in a similar manner as the antibiotic resistance gene products or a gene that expresses a product which acts like the herbicide target but is not affected by the herbicide. Herbicide tolerance may not only serve as a trait useful for selection in the development of transgenic plants, but also has some commercial interest. Herbicide tolerance transgenic plants are therefore among the first crops approaching market introduction.

Transformation of plant protoplasts, cells and tissues is usually only useful if the they can be regenerated into whole plants. The rates of regeneration vary greatly not only among different species, but also between cultivars of the same species. In many cases regeneration of whole plants from cells is not possible or very difficult. Besides the ability to introduce a gene into the genome of a plant species, regeneration of intact, fertile plants out of transformed cells or tissues is the most limiting step in developing transgenic plants.

CULTIGEN

A cultigen is a plant that has been deliberately altered or selected by humans; it is the result of artificial selection. These man-made or anthropogenic plants are, for the most part, plants with commercial value used in horticulture, agriculture or forestry. Because cultigens are defined by their mode of origin and not by where they are growing, plants meeting this definition remain cultigens whether they are naturalised in the wild, deliberately planted in the wild, or growing in cultivation. Cultigens arise in the following ways: selections of variants from the wild or cultivation including vegetative sports (aberrant growth that can be reproduced reliably in cultivation); plants that are the result of plant breeding and selection programs; genetically modified plants (plants modified by the deliberate implantation of genetic material); and graft-chimaeras (plants grafted to produce mixed tissue, the graft material possibly from wild plants, special selections, or hybrids).

Naming

Cultigens may be named in any of a number of way. The traditional method of scientific naming is under the International Code of Nomenclature for algae, fungi, and plants, and many of the most important cultigens, like maize (Zea mays) and banana (Musa acuminata), are so named. Although it is perfectly in order to give a cultigen a botanical name, in any rank desired, now or at any other time, these days it is more common for cultigens to be given names in accordance with the principles, rules and recommendations laid down in the International Code of Nomenclature for Cultivated Plants (ICNCP) which provides for the names of cultigens in three classification categories, the cultivar, the Group (formerly cultivar-group), and the grex. From that viewpoint it may be said that there is a separate discipline of cultivated plant taxonomy, which forms one of the ways to look at cultigens. The ICNCP does not recognize the use of trade designations and other marketing devices as scientifically acceptable names, but does provide advice on how they should be presented.

Not all cultigens have been given names according to the Cultivated Plant Code. Apart from ancient cultigens there may be occasional anthropogenic plants such as those that are the result of breeding, selection, and tissue grafting that are of no commercial value and have therefore not been given names according to the ICNCP.

The Wild/Cultivated Distinction

Interest in the distinction between wild and cultivated plants dates back to antiquity. Botanical historian Alan Morton notes that wild and cultivated plants (cultigens) were of intense interest to the ancient Greek botanists (partly for religious reasons) and that the distinction was discussed in some detail by Theophrastus the "Father of Botany". Theophrastus was a pupil of both Plato and Aristotle and succeeded the latter as head of the Peripatetic School of Philosophy at the Lyceum in Athens. Theophrastus accepted the view that it was human action, not divine intervention, that produced cultivated plants (cultigens) from wild plants and he also "had an inkling of the limits of culturally induced (phenotypic) changes and of the importance of genetic constitution". He also noted that cultivated varieties of fruit trees would degenerate if cultivated from seed.

Bailey's Definitions

Bailey soon altered his 1923 definition of cultigen when, in 1924, he gave a new definition in the Glossary of his Manual of Cultivated Plants as:

> "Plant or group known only in cultivation; presumably originating under domestication; contrast with indigen ".

This, in essence, is the definition given at the head of this piece. This definition of the cultigen permits the recognition of cultivars, unlike the 1923 definition which restricts the idea of the cultigen to plants at the rank of species.

In later publications of the Liberty Hyde Bailey Hortorium, Cornell, the idea of the cultigen having the rank of species returned both of these publications indicate that the terms cultigen and cultivar are not synonymous and that cultigens exist at the rank of species only.

> "A cultigen is a plant or group of apparent specific rank, known only in cultivation, with no determined nativity, presumably having originated, in the form in which we know it, under domestication. Compare indigen. Examples are Cucurbita maxima, Phaseolus vulgaris, Zea mays".

Recent usage in horticulture has, however, maintained a distinction between cultigen and cultivar while nevertheless allowing the inclusion of cultivars within the definition.

CULTIVAR

A cultivar (cultivated variety) is an assemblage of plants selected for desirable characteristics that are maintained during propagation. More generally, a cultivar is the most basic classification category of cultivated plants in the International Code of Nomenclature for Cultivated Plants (ICNCP). Most cultivars arose in cultivation, but a few are special selections from the wild.

Popular ornamental garden plants like roses, camellias, daffodils, rhododendrons, and azaleas are cultivars produced by careful breeding and selection for floral colour and form. Similarly, the world's agricultural food crops are almost exclusively cultivars that have been selected for characters such as improved yield, flavour, and resistance to disease, and very few wild plants are now used as food sources. Trees used in forestry are also special selections grown for their enhanced quality and yield of timber.

Cultivars form a major part of Liberty Hyde Bailey's broader group, the cultigen, which is defined as a plant whose origin or selection is primarily due to intentional human activity. A cultivar is not the same as a botanical variety, which is a taxonomic rank below subspecies, and there are differences in the rules for creating and using the names of botanical varieties and cultivars. In recent times, the naming of cultivars has been complicated by the use of statutory patents for plants and recognition of plant breeders' rights.

The International Union for the Protection of New Varieties of Plants (UPOV – French: Union internationale pour la protection des obtentions végétales) offers legal protection of plant cultivars to persons or organisations that introduce new cultivars to commerce. UPOV requires that a cultivar be "distinct, uniform", and "stable". To be "distinct", it must have characters that easily distinguish it from any other known cultivar. To be "uniform" and "stable", the cultivar must retain these characters in repeated propagation.

The naming of cultivars is an important aspect of cultivated plant taxonomy, and the correct naming of a cultivar is prescribed by the Rules and Recommendations of the International Code of Nomenclature for Cultivated Plants (ICNCP, commonly denominated the Cultivated Plant Code). A cultivar is given a cultivar name, which consists of the scientific Latin botanical name followed by a cultivar epithet. The cultivar epithet is usually in a vernacular language. For example, the full cultivar name of the King Edward potato is Solanum tuberosum 'King Edward'. 'King Edward' is the cultivar epithet, which, according to the Rules of the Cultivated Plant Code, is bounded by single quotation marks.

Cultigens

The words cultigen and cultivar may be confused with each other. A cultigen is any plant that is deliberately selected for or altered in cultivation, as opposed to an indigen; the Cultivated Plant Code states that cultigens are "maintained as recognisable entities solely by continued propagation". Cultigens can have names at any of many taxonomic ranks, including those of grex, species, cultivar group, variety, form, and cultivar; and they may be plants that have been altered in cultivation, including by genetic modification, but have not been formally denominated. A cultigen or a component of a cultigen can be accepted as a cultivar if it is recognisable and has stable characters. Therefore, all cultivars are cultigens, because they originate in cultivation, but not all cultigens are cultivars, because some cultigens have not been formally distinguished and denominated as cultivars.

Bread wheat, Triticum aestivum, is considered a cultigen, and is a distinct species from other wheats according to the biological species concept. Many different cultivars have been created within this cultigen. Many other cultigens are not considered to be distinct species, and can be denominated otherwise.

Different Kinds

A cultivar of the orchid genus Oncidium.

Which plants are chosen to be named as cultivars is simply a matter of convenience as the category was created to serve the practical needs of horticulture, agriculture, and forestry.

Members of a particular cultivar are not necessarily genetically identical. The Cultivated Plant Code emphasizes that different cultivated plants may be accepted as different cultivars, even if they have the same genome, while cultivated plants with different genomes may be regarded as the same cultivar. The production of cultivars generally entails considerable human involvement although in a few cases it may be as little as simply selecting variation from plants growing in the wild (whether by collecting growing tissue to propagate from or by gathering seed).

Cultivars generally occur as ornamentals and food crops: Malus 'Granny Smith' and Malus 'Red Delicious' are cultivars of apples propagated by cuttings or grafting, Lactuca 'Red Sails' and Lactuca 'Great Lakes' are lettuce cultivars propagated by seeds. Named cultivars of Hosta and Hemerocallis plants are cultivars produced by micropropagation or division.

Clones

Leucospermum 'Scarlet Ribbon'.

Cultivars that are produced asexually are genetically identical and known as clones; this includes plants propagated by division, layering, cuttings, grafts, and budding. The propagating material may be taken from a particular part of the plant, such as a lateral branch, or from a particular phase of the life cycle, such as a juvenile leaf, or from aberrant growth as occurs with witch's broom. Plants whose distinctive characters are derived from the presence of an intracellular organism may also form a cultivar provided the characters are reproduced reliably from generation to generation. Plants of the same chimera (which have mutant tissues close to normal tissue) or graft-chimeras (which have vegetative tissue from different kinds of plants and which originate by grafting) may also constitute a cultivar.

Seed-produced

Some cultivars "come true from seed", retaining their distinguishing characteristics when grown from seed. Such plants are termed a "variety", "selection" or "strain" but these are ambiguous and confusing words that are best avoided. In general, asexually propagated cultivars grown from seeds produce highly variable seedling plants, and should not be labelled with, or sold under, the parent cultivar's name.

Seed-raised cultivars may be produced by uncontrolled pollination when characteristics that are distinct, uniform and stable are passed from parents to progeny. Some are produced as "lines" that are produced by repeated self-fertilization or inbreeding or "multilines" that are made up of several closely related lines. Sometimes they are F_1 hybrids which are the result of a deliberate repeatable single cross between two pure lines. A few F_2 hybrid seed cultivars also exist, such as Achillea 'Summer Berries'.

Some cultivars are agamospermous plants, which retain their genetic composition and characteristics under reproduction. Occasionally cultivars are raised from seed of a specially selected provenance – for example the seed may be taken from plants that are resistant to a particular disease.

Genetically Modified

Genetically modified plants with characteristics resulting from the deliberate implantation of genetic material from a different germplasm may form a cultivar. However, the International Code of Nomenclature for Cultivated Plants notes, "In practice such an assemblage is often marketed from one or more lines or multilines that have been genetically modified. These lines or multilines often remain in a constant state of development which makes the naming of such an assemblage as a cultivar a futile exercise". However, retired transgenic varieties such as the Fish tomato, which are no longer being developed, do not run into this obstacle and can be given a cultivar name.

Cultivars may be selected because of a change in the ploidy level of a plant which may produce more desirable characteristics.

Cultivar Names

Viola 'Clear Crystals Apricot'.

Every unique cultivar has a unique name within its denomination class (which is almost always the genus). Names of cultivars are regulated by the International Code of Nomenclature for Cultivated Plants, and may be registered with an International Cultivar Registration Authority (ICRA). There are sometimes separate registration authorities for different plant types such as roses and camellias. In addition, cultivars may be associated with commercial marketing names referred to in the Cultivated Plant Code as "trade designations".

Presenting in Text

A cultivar name consists of a botanical name (of a genus, species, infraspecific taxon, interspecific hybrid or intergeneric hybrid) followed by a cultivar epithet. The cultivar epithet is enclosed by single quotes; it should not be italicized if the botanical name is italicized; and each of the words within the epithet is capitalized (with some permitted exceptions such as conjunctions). It is permissible to place a cultivar epithet after a common name provided the common name is botanically unambiguous. Cultivar epithets published before 1 January 1959 were often given a Latin form and can be readily confused with the specific epithets in botanical names; after that date, newly coined cultivar epithets must be in a modern vernacular language to distinguish them from botanical epithets.

Examples of correct text presentation:

- Cryptomeria japonica 'Elegans'.
- Chamaecyparis lawsoniana 'Aureomarginata'.
- Chamaecyparis lawsoniana 'Golden Wonder'.
- Pinus densiflora 'Akebono' Apple 'Sundown'.

Some incorrect text presentation examples:

- Cryptomeria japonica "Elegans" (double quotes are unacceptable).
- Berberis thunbergii cv. 'Crimson Pygmy' (this once-common usage is now unacceptable, as it is no longer correct to use "cv". in this context; Berberis thunbergii 'Crimson Pygmy' is correct).
- Rosa cv. 'Peace' (this is now incorrect for two reasons: firstly, the use of "cv".; secondly, "Peace" is a trade designation or "selling name" for the cultivar R. 'Madame A. Meilland' and should therefore be printed in a different typeface from the rest of the name, without quote marks, for example: Rosa Peace).

Where several very similar cultivars exist they can be associated into a Group (formerly Cultivar-group). As Group names are used with cultivar names it is necessary to understand their way of presentation. Group names are presented in normal type and the first letter of each word capitalised as for cultivars, but they are not placed in single quotes. When used in a name, the first letter of the word "Group" is itself capitalized.

Presenting in Text

- Brassica oleracea Capitata Group (the group of cultivars including all typical cabbages).
- Brassica oleracea Botrytis Group (the group of cultivars including all typical cauliflowers).
- Hydrangea macrophylla Groupe Hortensia (in French) = Hydrangea macrophylla Hortensia Group (in English).

Where cited with a cultivar name the group should be enclosed in parentheses, as follows:

Hydrangea macrophylla (Hortensia Group) 'Ayesha'.

COMPOSITE CROSS POPULATION

A composite cross population (CCP) is created by crossing a number of plants from different lines, and subsequently bulking seeds from the resulting offspring. This makes a CCP a population of plants with a lot of inherent genetic diversity, in contrast to monocultures where all plants are clones and homozygous at all loci (fully inbred). In recent years CCPs have been proposed as a way to create modern landraces of wheat, barley and oats. Research is done to explore whether they are better suited for organic farming than the modern cultivars. They are suited for participatory breeding of crops, which is in contrast to cultivars owned by big breeding companies.

The idea of using CCPs in plant breeding was initially published in 1956. Yield data for 4 different populations for 8–28 years were presented in the article and after 8–15 years of repeated breeding under natural selection, the populations out-yielded the reference cultivar.

Stages

Creating a CCP involves 3 steps: Initiation, multiplication and mixture. The population then goes into the maintenance phase.

Initiation

A number of lines, generally 7-30, with interesting properties, such as yield or baking quality, are selected and all possible crosses of them are done. If many lines of different genetic background are used, a huge amount of genetic diversity will be present.

Multiplication

Seeds from crosses are sown out and harvested separately for a growing season or two until enough seeds are available.

Mixture

All seeds are mixed in equal portions to produce the first CCP generation.

Maintenance

The population is grown repeatedly and possibly changes due to natural selection. Each year seeds are saved after harvest, and used as seed for the next growing season. Plants that are successful under the prevailing growing conditions will give more seeds and contribute more to the next generation, compared to less successful plants. Disease will cull susceptible plants and the population will over time become resistant to the common diseases, but only if the initial population has resistance genes present.

Pros

When genotypes grown together in a diverse population have different profiles of resource use they complement each other in the exploitation of the limiting resource and therefore are subject to smaller between-plant competition.

In case of disease or environmental change some plants will take over when others fail. Yield stability over years and environments can be better than pure lines due to compensation.

Participatory plant breeding (PPB) methods represent alternatives aimed to improve local adaptation breeding, to promote genetic diversity, to empower farmers and rural communities. In PPB farmers are actively participating in developing new cultivars or populations, e.g. by performing selection.

Cons

Different genotypes may compete with each other in a way that is not beneficial for the population. A pure line optimized for a certain environment or usage outperforms the CCP under these specific conditions.

There is no reason to believe that the Darwinian selection will work in the desired direction for desired traits such as baking quality.

Common Bunt is a seed borne disease in wheat. In conventional farming it is controlled by fungicide treatment of seeds. In organic farming seeds can be cleaned by brushing before sowing, but it is also desirable that plants have genetic resistance. A CCP, including crosses of resistant cultivars, was grown with heavy common bunt infection for 5 years and it appeared to get more resistant, but the common bunt's virulence appeared to change at least as fast. The overall result was that infection levels went up.

DOUBLE-PAIR MATING

Double-pair mating (DPM) is a mating (crossing) design used in plant breeding. Each individual is mated with two others.

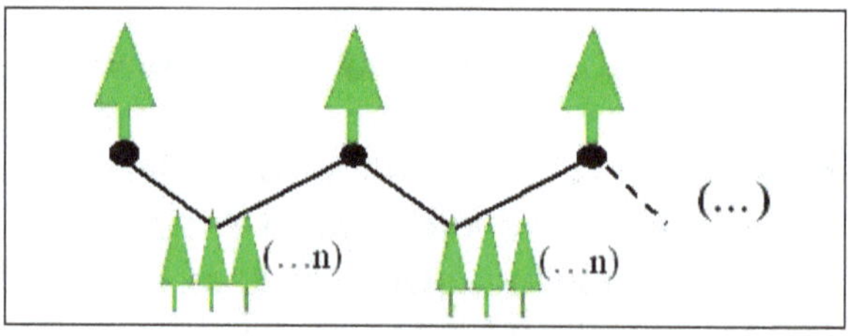

DMP scheme.

DPM is an efficient mating design in balanced breeding programmes, where equal contribution from each breeding population member is desired. With DPM the number of new families created is equal to the number of individuals mated. DPM allows to efficiently utilise positive assortative mating for more efficient use of the breeding population members for deployment to seed orchards. In comparison with single pair mating, DPM has the advantages that the genes from the individual are transmitted to next generation even if one of the crosses fails; that safer estimates of breeding values of the parents get possible (useful for seed orchards, where tested trees are preferred); and that genes from different ancestors have a better chance to combine.

NEW BREEDING TECHNIQUES

New Breeding Techniques (NBT), also named New Plant Engineering Techniques, are a suite of methods that could increase and accelerate the development of new traits in plant breeding. These new techniques, often involve 'genome editing' whose intention is to modify DNA at specific locations within the plants' genes so that new traits and properties are produced in crop plants.

An ongoing discussion in many countries is as to whether NBTs should be included within the same pre-existing governmental regulations to control genetic modification.

Methods Involved

New breeding techniques (NBTs) make specific changes within plant DNA in order to change its traits, and these modifications can vary in scale from altering single base, to inserting or removing one or more genes. The various methods of achieving these changes in traits include the following:

- Cutting and modifying the genome during the repair process (three tools are used to achieve this: Zinc finger nuclease; TALENs, and CRISPR/Cas Tools).
- Genome editing to introduce changes to just a few base pairs (using a technique called 'oligonucleotide-directed mutagenesis' (ODM)).
- Transferring a gene from an identical or closely related species (cisgenesis).
- Adding in a reshuffled set of regulatory instructions from same species (intra-genesis).
- Deploying processes that alter gene activity without altering the DNA itself (epi-genetic methods).
- Grafting of unaltered plant onto a genetically modified rootstock.

SELECTION METHODS IN PLANT BREEDING BASED ON MODE OF REPRODUCTION

Plant breeders use different methods depending on the mode of reproduction of crops, which include:

- Self-fertilization, where pollen from a plant will fertilise reproductive cells or ovules of the same plant.
- Cross-pollination, where pollen from one plant can only fertilize a different plant.
- Asexual propagation (e.g. runners from strawberry plants) where the new plant is genetically identical to its parent.
- Apomixis (self-cloning), where seeds are produced asexually and the new plant is genetically identical to its parent.

The mode of reproduction of a crop determines its genetic composition, which, in turn, is the deciding factor to develop suitable breeding and selection methods. Knowledge of mode of reproduction is also essential for its artificial manipulation to breed improved types. Only those breeding and selection methods are suitable for a crop which does not interfere with its natural state or ensure the maintenance of such a state. It is due to such reasons that imposition of self-fertilization on cross-pollinating crops leads to drastic reduction in their performance.

For teaching purpose, plant breeding is presented as four categories: Line breeding (autogamous crops), population breeding (allogamous crops), hybrid breeding (mostly allogamous crops, some autogamous crops), clone breeding (vegetatively propagated crops).

Self Fertilizing Crops (Autogamous Crops)

Certain restrictions caused the mechanisms for self-fertilization (partial and full self-fertilization) to develop in a number of plant species. Some of the reasons why a self-fertilizing method of reproduction is so effective are the efficacy of reproduction, as well as decreasing genetic variation and thus the fixation of highly adapted genotypes. Almost no inbreeding depression occurs in self-fertilizing plants because the mode of reproduction allows natural selection to take place in wild populations of such plants.

Critical steps in the improvement of self-fertilizing crops are the choice of parents and the identification of the best plants in segregating generations. The breeder should also have definite goals with the choice of parents. Self-fertilizing are easier to maintain, but this could lead to misuse of seed.

Some of the agronomy important, self-fertilizing crops include wheat, rice, barley, dry beans, soy beans, peanuts, tomatoes, etc.

Mass Selection

This method of selection depends mainly on the selection of plants according to their phenotype and performance. The seed from selected plants is bulked for the next generation. This method is used to improve the overall population by positive or negative mass selection. Mass selection is only applied to a limited degree in self-fertilizing plants and is an effective method for the improvement of land races. This method of selection will only be effective for highly heritable traits. One shortage of mass selection is the large influence that the environment has on the development, phenotype, and performance of single plants. This can also be an advantage in that varieties can be selected for local performance.

Stratified mass selection for ear size over 22 cycles has drastically altered plant phenotype in the maize population Zacatecas 58. Plants in the C22 cycle were 50 cm taller had twice the leaf area index, reached anthesis 7 days later and had a 30% higher harvest index than Co Differences in growth were detected early in ontogeny. The root growth of C22 exceeded that of Co and the ratio of shoot dry mass to root dry mass was reduced by nearly 12%, from 8.0±0.2 to 7.1±0.1. Analysis of yield components revealed that C22 was superior to Co in grain weight, number of rows per ear, number of grains per row, and total yield per unit area. Because the two genotypes were phenologically different, planting density optima are probably different for each population.

Selection of Cross-pollinated Crops

Plant species where normal mode of seed set is through a high degree of cross-pollination have characteristic reproductive features and population structure. Existence of self-sterility, self-incompatibility, imperfect flowers, and mechanical obstructions make the plant dependent upon foreign pollen for normal seed set. Each plant receives a blend of pollen from a large number of individuals each having different genotypes. Such populations are characterized by a high degree of heterozygosity with tremendous free and potential genetic variation, which is maintained in a steady state by free gene flow among individuals within the populations.

In the development of hybrid varieties, the aim is to identify the most productive heterozygote from the population, which then is produced with the exclusion of other members of the population.

Mass Selection

It is the simplest, easiest and oldest method of selection where individual plants are selected based on their phenotypic performance, and bulk seed selection proved to be quite effective in maize improvement at the initial stages but its efficacy especially for improvement of yield, soon came under severe criticism that culminated in the refinement of the method of mass selection. The selection after pollination does not provide any control over the pollen parent as result of which effective selection is limited only to female parents. The heritability estimates are reduced by half, since only parents are used to harvest seed whereas the pollen source is not known after the cross pollination has taken place.

Recurrent Selection

This type of selection is a refined version of the mass selection procedure and differs as follows:

- Visually selected individuals out of the base population undergo progeny testing.
- Individuals selected on basis of the progeny test data are crossed with each other in every possible way to produce seed to form the new base population.

Half-sib Selection with Progeny Testing

Selections are made based on progeny test performance instead of phenotypic appearance of the parental plants. Seed from selected half-sibs, which have been pollinated by random pollen from the population (meaning that only the female parent is known and selected, hence the term "half-sib") is grown in unreplicated progeny rows for the purpose of selection. A part of the seed is planted to determine the yielding ability, or breeding value, for any character of each plant. The seed from the most productive rows or remnant seed from the outstanding half-sibs is bulked to complete one cycle of selection.

Full-sib Selection with Progeny Testing

A number of full-sib families, each produced by making crosses between the two plants from the base population are evaluated in replicated trials. A part of each full-sib family is saved for recombination. Based on evaluation the remnant seed of selected full-sib families is used to recombine the best families.

Breeding of Asexually Propagated Crops

Asexual reproduction covers all those modes of multiplication of plants where normal gamete formation and fertilization does not take place making these distinctly different from normal seed production crops. In the absence of sexual reproduction, the genetic composition of plant material being multiplied remains essentially the same as its source plant.

Clones of mother plants can be made with the exact genetic composition of the mother plant. Superior plants are selected and propagated vegetatively; the vegetative propagated offspring are used to develop stable varieties without any deterioration due to segregation of gene combinations. This unique characteristic of asexual reproduction helped to develop a number of cultivars of fruits and vegetables including grapes, apples, pears and peaches.

Improving Asexual Plant Material through Selection

The selection in these crops is restricted to the material introduced from other sources, such as field plantations. The improvement of asexually propagated plants through induced mutations has distinct advantages and limitations. Any vegetative propagule can be treated with mutagens and even a single desirable mutant or a part of a mutated propagule (chimera) can be multiplied as an improved type of the original variety.

Selection of Asexual Plants

Selection, in the case of asexual plants, can be defined as the selection of the best performing plant and the vegetative propagation thereof. Because plants are not totally genetically stable, it can be expected that deviations would occur through the years. Selection is thus an ongoing process where deviants are selected or removed from the selection program. The main purpose of selection is to better the quality and yield of forthcoming plantations. Different approaches can be followed in the selection process of asexual plants, such as mass selection and clone selection from clone blocks.

In mass selection there are some factors that must be considered when selecting plants in a mother block, e.g. vineyard. Time of selection is a big factor, because you have to select when most of the characteristics of the plant are clearly showing. With asexual perennials the best time is just before harvest. For the best results the selected plant must be evaluated during the next season, when growth-abnormalities, leave disfigurations and virus symptoms are best visualized. Mass selection is done annually on the same plant for a minimum of three years. A plant that does not conform to the requirements in any given year of the selection cycle is discarded from the program.

New Clone Development

The development and registration of new clones take place by means of local clone selection in old plantations, as well as the importation of high quality clones from abroad, for local evaluation.

A clone is the vegetative offspring of one specific mother plant; it does not show any genetic, morphologic or physiologic deviations from the mother plant. Evaluation takes place with the different selected clones after selection.

BREEDING FOR HEAT STRESS TOLERANCE

Plant breeding is process of development of new cultivars. Plant breeding involves development of varieties for different environmental conditions – some of them are not favorable. Among them, heat stress is one of such factor that reduces the production and quality significantly. So breeding against heat is a very important criterion for breeding for current as well as future environments produced by global climate change (e.g. global warming).

Breeding for Heat Stress Tolerance in Plants

Heat stress due to increased temperature is a very important problem globally. Occasional or prolonged high temperatures cause different morpho-anatomical, physiological and biochemical changes in plants. The ultimate effect is on plant growth as well as development and reduced yield and quality. Breeding for heat stress tolerance can be mitigated by breeding plant varieties that have improved levels of thermo-tolerance using different conventional or advanced genetic tools. Marker assisted selection techniques for breeding are highly useful. Recently 41 polymorphic SSR markers has been identified between a heat tolerant rice variety 'N22' and heat susceptible-high yielding variety 'Uma' for the development of new 'high yielding-heat tolerant' rice varieties.

What is Heat Stress Tolerance?

Heat stress is defined as increased temperature level sufficient to cause irreversible damage to plant growth and development. Generally a temperature rise, above usually 10 to 15 °C above ambient, can be considered heat shock or heat stress. Heat tolerance is broadly defined as the ability of the plant tolerate heat – means that grow and produce economic yield under high temperatures.

Significance: Current and Future - Global Warming

Heat stress is a serious threat to crop production globally Global warming is particularly consequence of increased level of green house gases such as CO_2, methane, chlorofluorocarbons and nitrous oxides. The Intergovernmental Panel on Climatic Change (IPCC) has predicted a rise of 0.3 °C per decade reaching to approximately 1 and 3 °C above the present value by 2025 and 2100 AD, respectively.

Physiological Consequence of Heat Stress

At very high temperatures cause severe cellular injury and cell death may occur within short time, thus leading to a catastrophic collapse of cellular organization. However, under moderately high temperatures, the injury can only occur after longer exposure to such a temperature however the plant efficiency can be severely affected. High temperature directly affect injuries such as protein denaturation and aggregation, and increased fluidity of membrane lipids. Other indirect or slower heat injuries involve inactivation of enzymes in chloroplast and mitochondria, protein degradation, inhibition of protein synthesis, and loss of membrane integrity. Heat stress associated injuries ultimately lead to starvation, inhibition of growth, reduced ion flux, production of toxic compounds and production of reactive oxygen species (ROS). Immediately after exposure to high temperature stress-related proteins are expressed as stress defense strategy of the cell. Expression of heat shock proteins (HSPs), protein with 10 to 200 kDa, is supposed to be involved in signal transduction during heat stress. In many species it has been demonstrated that HSPs results in improved physiological phenomena such as photosynthesis, assimilate partitioning, water and nutrient use efficiency, and membrane stability.

Studies have found tremendous variation within and between species, thus this will help to breed heat tolerance for future environment. Some of attempts to develop heat-tolerant genotypes are successful.

Traits Associated with Heat Stress Tolerance

Different physiological mechanisms may contribute to heat tolerance in the field for example, heat tolerant metabolism as indicated by higher photosynthetic rates, stay-green, and membrane thermo-stability, or heat avoidance as indicated by canopy temperature depression. Several physiological and morphological traits have been evaluated for heat tolerance - Canopy temperature, leaf chlorophyll, stay green, leaf conductance, spike number, biomass, and flowering date.

Canopy Temperature Depression

CTD has shown clear association with yield in warm environments shows it association with heat stress tolerance. CTD shows high genetic correlation with yield and high values of proportion of direct response to selection indicating that the trait is heritable and therefore amenable to early generation selection. Since an integrated CTD value can be measured almost instantaneously on scores of plants in a small breeding plot (thus reducing error normally associated with traits measured on individual plants), work has been conducted to evaluate its potential as an indirect selection criterion for genetic gains in yield. CTD is affected by many physiological factors, which makes it a powerful.

Stomatal Conductance

Canopy temperature depression is highly suitable for selecting physiologically superior lines in warm, low relative humidity environments where high evaporative demand leads to leaf cooling of up to 10 °C below ambient temperatures. This permits differences among genotypes to be detected relatively easily using infrared thermometry. However, such differences cannot be detected in high relative humidity environments because the effect of evaporative cooling of leaves is negligible. Nonetheless, leaves maintain their stomata open to permit the uptake of CO_2, and differences in the rate of CO_2 fixation may lead to differences in leaf conductance that can be measured using a porometer. Porometry can be used to screen individual plants. The heritability of stomatal conductance is reasonably high, with reported values typically in the range of 0.5 to 0.8. Plants can be assessed for leaf conductance using a viscous flow porometer that is available on the market (Thermoline and CSIRO, Australia). This instrument can give a relative measure of stomatal conductance in a few seconds, making it possible to identify physiologically superior genotypes from within bulks.

Membrane Thermostability

Although resistance to high temperatures involves several complex tolerance and avoidance mechanisms, the membrane is thought to be a site of primary physiological injury by heat, and measurement of solute leakage from tissue can be used to estimate damage to membranes. Since membrane thermostability is reasonably heritable and shows high genetic correlation with yield.

Chlorophyll Fluorescence

Chlorophyll fluorescence, an indication of the fate of excitation energy in the photosynthetic apparatus, has been used indicator for heat stress tolerance.

Chlorophyll Content and Stay Green

Chlorophyll content and stay green traits have been found to be associated with heat stress tolerance. Xu et al. identified three QTLs for chlorophyll content (coincided with three stay-green QTL regions) were identified in Sorghum. The Stg1 and Stg2 regions also contain the genes for key photosynthetic enzymes, heat shock proteins, and an abscisic acid (ABA) responsive gene.

Photosynthesis

Decliend photosynthesis is suggested as measure of heat stress sensitivity in plants.

Stem Reserve Remobilization

Combination Breeding and Physiological Breeding

The physiological-trait-based breeding approach has merit over breeding for yield per se because it increases the probability of crosses resulting in additive gene action. The concept of combination phenomics comes from the idea that two or more stress have common physiological effect or common traits - which is an indicator of overall plant health. Similar analogy in human medical terms is high blood pressure or high body temperature or high white blood cells in body is an indicator of health problems and thus we can select healthy people from unhealthy using such a measure. As both abiotic and abiotic stresses can result in similar physiological consequence, tolerant plant can be separated from sensitive plants. Some imaging or infrared measuring techniques can help to speed the process for breeding process. For example, spot blotch intensity and canopy temperature depression can be monitored with canopy temperature depression.

BREEDING FOR DROUGHT STRESS TOLERANCE

Breeding for drought resistance is the process of breeding plants with the goal of reducing the impact of dehydration on plant growth.

Dehydration Stress

Crop Plants

In nature or crop fields, water is often the most limiting factor for plant growth. If plants do not receive adequate rainfall or irrigation, the resulting dehydration stress can reduce growth more than all other environmental stresses combined.

Drought can be defined as the absence of rainfall or irrigation for a period of time sufficient to deplete soil moisture and cause dehydration in plant tissues. Dehydration stress results when water loss from the plant exceeds the ability of the plant's roots to absorb water and when the plant's water content is reduced enough to interfere with normal plant processes.

Global Phenomenon

About 15 million km2 of the land surface is covered by crop-land, and about 16% of this area is equipped for irrigation. Thus, in many parts of the world, including the United States, plants may frequently encounter dehydration stress. Rainfall is very seasonal and periodic drought occurs regularly. The effect of drought is more prominent in sandy soils with low water holding capacity. On such soils some plants may experience dehydration stress after only a few days without water.

During the 20th century, the rate of increase in 'blue' water withdrawal (from rivers, lakes, and aquifers) for irrigation and other purposes was higher than the growth rate of the world population. Country-wise maps of irrigated areas are available.

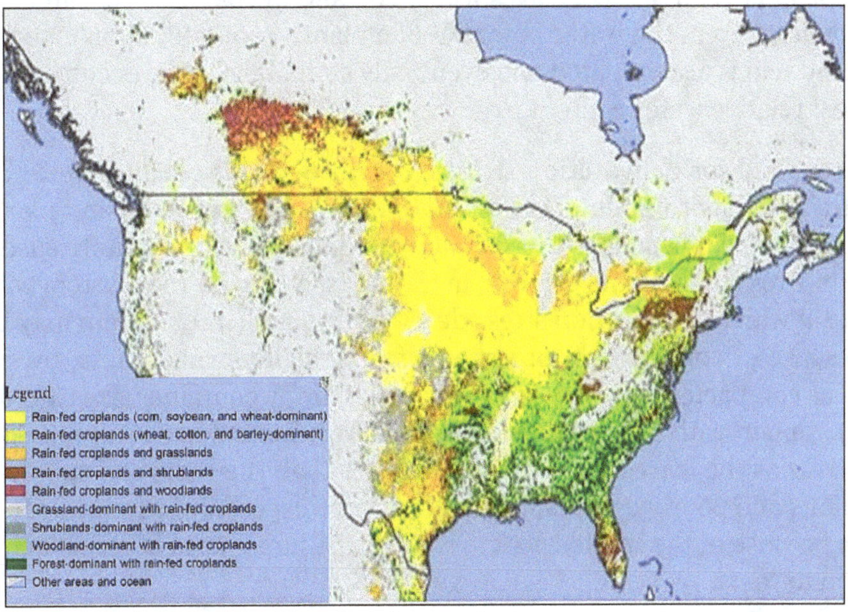

Rain fed areas of USA.

Future Challenges to Crop Production

Soil moisture deficit is a significant challenge to the future of crop production. Severe drought in parts of the U.S., Australia, and Africa in recent years drastically reduced crop yields and disrupted regional economies. Even in average years, however, many agricultural regions, including the U.S. Great Plains, suffer from chronic soil moisture deficits. Cereal crops typically attain only about 25% of their potential yield due to the effects of environmental stress, with dehydration stress the most important cause. Two major trends will likely increase the frequency and severity of soil moisture deficits:

- Climate change: Higher temperatures are likely to increase crop water use due to increased transpiration. A warmer atmosphere will also speed up melting of mountain snow pack, resulting in less water available for irrigation. More extreme weather patterns will increase the frequency of drought in some regions.
- Limited water supplies: Increased demand from municipal and industrial users will further reduce the amount of water available for irrigated crops.

Although changes in tillage and irrigation practices can improve production by conserving water, enhancing the genetic tolerance of crops to drought stress is considered an essential strategy for addressing moisture deficits.

Plant Physiology

A plant responds to a lack of water by halting growth and reducing photosynthesis and other plant processes in order to reduce water use. As water loss progresses, leaves of some species may appear to change colour usually to blue-green. Foliage begins to wilt and, if the plant is not irrigated, leaves will fall off and the plant will eventually die. Soil moisture deficit lowers the water potential of a plant's root and, upon extended exposure, abscisic acid is accumulated and eventually stomatal closure occurs. This reduces a plant's leaf relative water content.

The time required for dehydration stress to occur depends on the water-holding capacity of the soil, environmental conditions, stage of plant growth, and plant species. Plants growing in sandy soils with low water-holding capacity are more susceptible to dehydration stress than plants growing in clay soils. A limited root system will accelerate the rate at which dehydration stress develops. A plant's root system may be limited by the presence of competing root systems from neighbouring plants, by site conditions such as compacted soils or high water tables, or by container size (if growing in a container). A plant with a large mass of leaves in relation to the root system is prone to drought stress as the leaves may lose water faster than the roots can supply it. Newly planted plants and poorly established plants may be especially susceptible to dehydration stress because of the limited root system or the large mass of stems and leaves in comparison to roots.

Other Stress Factors

Aside from the moisture content of the soil, environmental conditions of high light intensity, high temperature, low relative humidity and high wind speed will significantly increase plant water loss. The prior environment of a plant also can influence the development of dehydration stress. A plant that has been exposed to dehydration stress (hardened) previously and has recovered may become more drought resistant. Also, a plant that was well-watered prior to being water-limited will usually survive a period of drought better than a continuously dehydration-stressed plant.

Mechanisms of Drought Resistance

The degree of resistance to drought depends upon individual crops. Generally three strategies can help a crop to mitigate the effect of dehydration stress:

Avoidance

If the plant shows dehydration avoidance, the environmental factor is excluded from the plant tissues by reducing water loss ("water savers", e.g. by thick leaf epicuticular wax, leaf rolling, leaf posture) or maintaining water uptake ("water spenders", e.g. by growing deeper roots). Dehydration avoidance is desirable in modern agriculture, where drought resistance requires the maintenance of economically viable plant production under dehydration stress. The role of dehydration avoidance is maintaining water supply and sustaining leaf hydration and turgidity with the purpose of maintaining stomatal opening and transpiration as long as possible under water deficit. This is essential for leaf gas exchange, photosynthesis and plant production through carbon assimilation.

Tolerance

If the plant shows dehydration tolerance, the environmental factor enters the plant tissues but the tissues survive, by e.g. maintaining turgor and osmotic adjustment.

Escape

Dehydration escape involves e.g. early maturing or seed dormancy, where the plant uses previous optimal conditions to develop vigor. Dehydration recovery refers to some plant species being able to recuperate after brief drought periods.

A proper timing of life-cycle, resulting in the completion of the most sensitive developmental stages while water is abundant, is considered to be a dehydration escape strategy. Avoiding dehydration stress with a root system capable of extracting water from deep soil layers, or by reducing evapotranspiration without affecting yields, is considered as dehydration avoidance. Mechanisms such as osmotic adjustment (OA) whereby a plant maintains cell turgor pressure under reduced soil water potential are categorised as dehydration tolerance mechanisms. Dehydration avoidance mechanisms can be expressed even in the absence of stress and are then considered constitutive. Dehydration tolerance mechanisms are the result of a response triggered by dehydration stress itself and are therefore considered adaptive. When the stress is terminal and predictable, dehydration escape through the use of shorter duration varieties is often the preferable method of improving yield potential. Dehydration avoidance and tolerance mechanisms are required in situations where the timing of drought is mostly unpredictable.

Drought resistance mechanisms are genetically controlled and genes or QTL responsible for drought resistance have been discovered in several crops which opens avenue for molecular breeding for drought resistance.

Drought Resistance Traits

Resistance to drought is a quantitative trait, with a complex phenotype, often confounded by plant phenology. Breeding for drought resistance is further complicated since several types of abiotic stress, such as high temperatures, high irradiance, and nutrient toxicities or deficiencies can challenge crop plants simultaneously.

Osmotic Adjustment

When a plant is exposed to water deficit, it may accumulate a variety of osmotically active compounds such as amino acids and sugars, resulting in a lowering of the osmotic potential. Examples of amino acids that may be up-regulated are proline and glycine betaine. This is termed osmotic adjustment and enables the plant to take up water, maintain turgor and survive longer.

Cell Membrane Stability

The ability to survive dehydration is influenced by a cell's ability to survive at reduced water content. This can be considered complementary to OA because both traits will help maintain leaf growth (or prevent leaf death) during water deficit. Crop varieties differ in dehydration tolerance and an important factor for such differences is the capacity of the cell membrane to prevent electrolyte leakage at decreasing water content, or "cell membrane stability (CMS)". The maintenance of membrane function is assumed to mean that cell activity is also maintained. Measurements of CMS have been used in different crops and are known to be correlated with yields under high temperature and possibly under dehydration stress.

Epicuticular Wax

In sorghum (Sorghum bicolor L. Moench), drought resistance is a trait that is highly correlated with the thickness of the epicuticular wax layer. Experiments have demonstrated that rice varieties with a thick cuticle layer retain their leaf turgor for longer periods of time after the onset of a water-stress.

Partitioning and Stem Reserve Mobilisation

As photosynthesis becomes inhibited by dehydration, the grain filling process becomes increasingly reliant on stem reserve utilisation. Numerous studies have reported that stem reserve mobilisation capacity is related to yield under dehydration stress in wheat. In rice, a few studies also indicated that this mechanism maintains grain yield under dehydration stress at the grain filling stage. This dehydration tolerance mechanism is stimulated by a decrease in gibberellic acid concentration and an increase in abscisic acid concentration.

Seedling Drought Resistance Traits

For emergence from deep sowing (to exploit dry upper soil), this is practised to help seedlings reach the receding moisture profile, and to avoid high soil surface temperatures which inhibit germination. Screening at these stage provides practical advantages, specially when managing large amount of germ-plasms.

The Drought Resistant Ideotype

Usually ideotypes are developed to create an ideal plant variety. The following traits constitutes ideotype of wheat by CIMMYT:

- Large seed size:Helps emergence, early ground cover and initial biomass.
- Long coleoptiles: For emergence from deep sowing.
- Early ground cover.

Thinner, wider leaves (i.e., with a relatively low specific leaf weight) and a more prostrate growth habit help to increase ground cover, thus conserving soil moisture and potentially increasing radiation use efficiency.

- High pre-anthesis biomass.
- Good capacity for stem reserves and remobilisation.
- High spike photosynthetic capacity.
- High RLWC/GS/CTD during grain filling to indicate ability to extract water.
- Osmotic adjustment.
- Accumulation of ABA.

The benefit of ABA accumulation under dehydration has been demonstrated. It appears to pre-adapt plants to stress by reducing stomatal conductance, rates of cell division, organ size, and increasing development rate. However, high ABA can also result in sterility problems since high ABA levels may abort developing florets:

- Heat Tolerance: The contribution of heat tolerance to performance under dehydration stress needs to be quantified, but it is relatively easy to screen for.
- Leaf anatomy: Waxiness, pubescence, rolling, thickness, posture. These traits decrease radiation load to the leaf surface. Benefits include a lower evapotranspiration rate and reduced risk of irreversible photo-inhibition. However, they may also be associated with reduce radiation use efficiency, which would reduce yield under more favourable conditions.
- High tiller survival: Comparison of old and new varieties have shown that under dehydration older varieties over-produce tillers many of which fail to set grain while modern drought resistant lines produce fewer tillers most of which survive.
- Stay-green: The trait may indicate the presence of drought resistance mechanisms, but probably does not contribute to yield per se if there is no water left in the soil profile by the end of the cycle to support leaf gas exchange. It

may be detrimental if it indicates lack of ability to remobilise stem reserves. However, research in sorghum has indicated that Stay-green is associated with higher leaf chlorophyll content at all stages of development and both were associated with improved yield and transpiration efficiency under dehydration.

Combination Phenomics: Overall Health of Crops

The concept of combination phenomics comes from the idea that two or more plant stresses have common physiological effects or common traits - which are an indicator of overall plant health. As both biotic and abiotic stresses can result in similar physiological consequence, drought resistant plants can be separated from sensitive plants. Some imaging or infrared measuring techniques can help to speed the process for breeding process. For example, spot blotch intensity and canopy temperature depression can be monitored with canopy temperature depression.

Molecular Breeding for Drought Resistance:Recent research breakthroughs in biotechnology have revived interest in targeted drought resistance breeding and use of new genomics tools to enhance crop water productivity. Marker-assisted breeding is revolutionising the improvement of temperate field crops and will have similar impacts on breeding of tropical crops. Other molecular breeding tool include development of genetically modified crops that can tolerate plant stress. As a complement to the recent rapid progress in genomics, a better understanding of physiological mechanisms of dehydration response will also contribute to the progress of genetic enhancement of crop drought resistance. It is now well accepted that the complexity of the dehydration syndrome can only be tackled with a holistic approach that integrates physiological dissection of crop dehydration avoidance and - tolerance traits using molecular genetic tools such as MAS, micro-arrays and transgenic crops, with agronomic practices that lead to better conservation and utilisation of soil moisture, and better matching of crop genotypes with the environment. MAS has been implemented in rice varieties to assess the drought tolerance and to develop new abiotic stress-tolerant varieties.

HYBRIDIZATION

Hybridization is the process of interbreeding between individuals of different species (interspecific hybridization) or genetically divergent individuals from the same species (intraspecific hybridization). Offspring produced by hybridization may be fertile, partially fertile, or sterile.

Crop yields increase dramatically when hybridization is used to exceed one or more of the parents in size and reproductive potential.

Plants hybridize much more frequently and successfully than animals do. Pollen from flowering plants disperses widely and may land on flowers of other species. Chromosomal doubling (polyploidy) occurs more frequently in plants and facilitates the fertility of the hybrid offspring. Finally, plant forms are less stringently controlled than animal forms, and so the intermediate form of a plant hybrid is more likely to be physiologically successful.

One of the first persons to study plant hybridization was Josef Kölrueter, who published the results of his experiments on tobacco in 1760. Kölrueter concluded that interspecific hybridization in nature is rare unless humans disturb the habitat. Since that time, many instances of hybridization among various plant species have been documented.

One good example of plant hybridization involves hybridization between the elegant sego lily (Calochortus selwayensis) and a mariposa lily (C. apiculatus) in western Montana. The sego lily, with purple-spotted petals, lives in dry sites at mid-elevations in the Rocky Mountains under the somewhat open canopy of ponderosa pine forests. The mariposa, with its cream-colored petals, lives in moister sites at higher elevations under the more closed Douglas-fir canopies. Interspecific hybrids between the elegant sego and mariposa lilies are found in great abundance on ski slopes where Douglas-fir canopies have been opened and kept clear of trees and tall shrubs.

The ski slope is a habitat that is too dry and too open for the mariposa to thrive and too moist for the elegant sego, but just right for the hybrids. Such an intermediate habitat is called a hybrid habitat. Human disturbance can generate hybrid habitats of many types and thus can foster interspecific hybridization.

Backcrossing, which is the interbreeding between hybrids and their parental species, can transfer alleles from one parent to the other using the interspecific hybrids as a genetic bridge in a process called introgression. Introgression increases the genetic variation of one or both of the parents. In the previous example, there is extensive backcrossing between the hybrids and mariposa lily. The result of this introgression is that some mariposa lilies now display petals with some of the purple spotting characteristic of the elegant sego, and theoretically they can live in slightly drier habitats.

Often interspecific hybrids are sterile or for some other reason cannot interbreed with the parental species. Occasionally sterile interspecific hybrids can undergo a doubling of their chromosome set and become fertile tetraploids (four sets of chromosomes). For example, the bread wheats that humans use today are a result of two hybridizations each followed by chromosome doubling to produce fertile hexaploids (six sets of chromosomes). In such instances the hybrids can become new species with characteristics different from either of the parents.

Humans have used intraspecific hybridization, hybridization between strains of a single species, to develop high-yielding crops. In corn, continually inbred varieties will often exhibit inbreeding depression, which is a reduction in vitality and yield. Hybridization between inbred lines can result in hybrids that exceed one or more of the parents in size and reproductive potential. This increased vitality is called hybrid vigor and has been studied since the time of English naturalist Charles Darwin. Crop yields increase dramatically (as much as 100 percent) when hybridization is used in this way. In the twenty-first century, over 90 percent of the corn grown is of hybrid origin.

References

- Plant-breeding-benefits-your-farm, farmers: seedgrowers.ca, Retrieved 18 January, 2020
- Spencer, Roger D.; Cross, Robert G. (2007). "The International Code of Botanical Nomenclature (ICBN), the International Code of Nomenclature for Cultivated Plants (ICNCP), and the cultigen". Taxon. 56 (3): 938–940. doi:10.2307/25065875. JSTOR 25065875
- Hybridization-Plant: biologyreference.com, Retrieved 16 March, 2020
- Döring, Thomas F.; Knapp, Samuel; Kovacs, Geza; Murphy, Kevin; Wolfe, Martin S. (17 October 2011). "Evolutionary Plant Breeding in Cereals—Into a New Era". Sustainability. 2011 (3). doi:10.3390/su3101944. Retrieved 20 January 2018
- Lusser M, Parisi C, Rodriguez Cerezo E, Plan D (18 October 2011). "New plant breeding techniques. State-of-the-art and prospects for commercial development. Publications Office of the European Union". doi:10.2791/54761

PERMISSIONS

All chapters in this book are published with permission under the Creative Commons Attribution Share Alike License or equivalent. Every chapter published in this book has been scrutinized by our experts. Their significance has been extensively debated. The topics covered herein carry significant information for a comprehensive understanding. They may even be implemented as practical applications or may be referred to as a beginning point for further studies.

We would like to thank the editorial team for lending their expertise to make the book truly unique. They have played a crucial role in the development of this book. Without their invaluable contributions this book wouldn't have been possible. They have made vital efforts to compile up to date information on the varied aspects of this subject to make this book a valuable addition to the collection of many professionals and students.

This book was conceptualized with the vision of imparting up-to-date and integrated information in this field. To ensure the same, a matchless editorial board was set up. Every individual on the board went through rigorous rounds of assessment to prove their worth. After which they invested a large part of their time researching and compiling the most relevant data for our readers.

The editorial board has been involved in producing this book since its inception. They have spent rigorous hours researching and exploring the diverse topics which have resulted in the successful publishing of this book. They have passed on their knowledge of decades through this book. To expedite this challenging task, the publisher supported the team at every step. A small team of assistant editors was also appointed to further simplify the editing procedure and attain best results for the readers.

Apart from the editorial board, the designing team has also invested a significant amount of their time in understanding the subject and creating the most relevant covers. They scrutinized every image to scout for the most suitable representation of the subject and create an appropriate cover for the book.

The publishing team has been an ardent support to the editorial, designing and production team. Their endless efforts to recruit the best for this project, has resulted in the accomplishment of this book. They are a veteran in the field of academics and their pool of knowledge is as vast as their experience in printing. Their expertise and guidance has proved useful at every step. Their uncompromising quality standards have made this book an exceptional effort. Their encouragement from time to time has been an inspiration for everyone.

The publisher and the editorial board hope that this book will prove to be a valuable piece of knowledge for students, practitioners and scholars across the globe.

INDEX

www.ingramcontent.com/pod-product-compliance
Lightning Source LLC
LaVergne TN
LVHW080612210926
848485LV00003B/142

* 9 7 8 1 6 3 9 8 9 4 2 1 5 *